Public Power

Leah,

People like you inspired this book. Thank you for caring about the public services we all need.

Public Power

THE FIGHT FOR PUBLICLY OWNED ELECTRICITY

Howard Hampton
with Bill Reno

INSOMNIAC PRESS

Copyright © 2003 by Howard Hampton.

All rights reserved. No part of this publication may be reproduced, stored in a retrieval system or transmitted, in any form or by any means, without the prior written permission of the publisher or, in the case of photocopying or other reprographic copying, a license from Access Copyright (Canadian Copyright Licensing Agency), 1 Yonge St., Suite 1900, Toronto, Ontario, Canada, M5E 1E5.

Edited and designed by Mike O'Connor
Copy-edited by Richard Almonte

National Library of Canada Cataloguing in Publication Data

Hampton, Howard
Public power: the fight for publicly owned electricity / Howard Hampton.

Includes bibliographical references.
ISBN 1-894663-44-6

1. Electric utilities—Government ownership—Canada. 2. Electric utilities—Government ownership—United States. 3. Privatization—Canada. 4. Privatization--United States. I. Title.

HD9685.N662H34 2003   333.793'2'0971   C2003-900715-4

The publisher gratefully acknowledges the support of the Canada Council, the Ontario Arts Council and the Department of Canadian Heritage through the Book Publishing Industry Development Program.

Printed and bound in Canada.

Insomniac Press, 192 Spadina Avenue, Suite 403,
Toronto, Ontario, Canada, M5T 2C2
www.insomniacpress.com

To our children and those who come after us,
may we leave you a better world.

Acknowledgments

I am grateful to a great many people who helped me put together this book. First and foremost, I want to thank Bill Reno, without whose keen insight, hard work and collaborative writing assistance, this book would not have been possible. As well, I want to thank Richard Almonte and Mike O'Connor for their excellent editing.

To former NDP Leader Donald C. MacDonald, a true public power visionary, thank you for your fierce and tireless advocacy on behalf of the people of Ontario. Your ideas on energy were ahead of their time and current events bear out how right you were and how crucial the battle was that you waged.

I am also indebted for the help of Anne Wordsworth, who reviewed parts of the book and gave excellent advice, and to Brian Charlton and Ruth Grier, whose wise counsel and recollections of the NDP's years in opposition and government were invaluable.

John Cook and the crew at Compass 360 came up with the cover design. Dommi Freestone, the Corporate Archives Assistant at the Hydro One Archives, was extremely generous with her time and went out of her way to help us find many of this book's excellent photographs. I encountered the same warm response from the City of Toronto Archives and I'm grateful for that as well. The extraordinary staff at the Queen's Park Legislative Library helped me track down every detail I needed to write this book with the same customary patience and tolerance they've shown me since my first election in 1987.

I am indebted to my hard-working staff and NDP Caucus colleagues who took time, over and above their daily duties, to help put together this book – especially Rob Milling, Sheila White, Fred Gloger, Charles Campbell, Gil Hardy, Ezia Cervoni, Marcia McVea, Greg Bennett, Stephanie Levesque, Elliott Anderson, Jacob Leibovitch, Jeffrey Ferrier and Graham Mitchell. And thank you to Joan Milling for the 101 persnickety tasks and details she handled with endless grace and efficiency.

Special thanks go to all the Hydro Hotheads across Ontario for helping New Democrats put hydro on the political radar screen. Whether you met the Public Power Bus, circulated or signed petitions, wrote letters to your local newspapers, hosted town hall meetings or burned your hydro bills, you fuelled the NDP's campaign to stop hydro privatization and deregulation. And you moved me to write this book. Paul Kahnert and the Ontario Electricity Coalition were equally inspiring to me and I must thank them for their tireless grassroots campaign that resonated across this province. I, indeed all Ontarians, owe a deep debt of gratitude to CEP and CUPE for launching the important legal challenge that derailed the wholesale privatization of Hydro One.

A big "thank-you" to all the hydro workers, utility board members and staff I've worked with or spoken to over the years for sharing with me their interest, information and knowledge of hydro. Interacting with those many people piqued my interest in the history of hydro and I hope with this offering I have done justice to their spirit and passion for the subject.

Finally, on a personal note, I want to thank my long-suffering parents, George and Elsie Hampton, who support me in just about everything I do. Thanks also to the hard-working people of Kenora-Rainy River, who continue to inspire me with their confidence and trust. Finally, and most importantly, thanks to my Number One supporters, Shelley, Sarah and Jonathan, who unselfishly allowed me the time to get this done and who put up with the late nights and long hours involved.

Contents

Introduction   15
One hundred years later, public vs. private power is once again a dominant economic and political issue. Will the outcome be different this time?

1. The White Coal of Niagara   21
At the dawn of the electricity age, private interests came very close to long term monopoly control over Niagara's immense hydroelectric potential.

2. The Triumph of Democracy   27
Failing to respond to growing popular demand for public power, the Ross Liberal government was crushed at the polls by the Whitney Conservatives.

3. The People's Power Company   39
Led by the remarkable Adam Beck, public power forces dramatically defeated private power interests in two municipal plebiscites, in the courts, and at Queen's Park.

4. Rural Electrification, Farm Modernization   55
It took decades to fulfill Beck's vision of bringing affordable power to all of Ontario's vast rural regions. The campaign began in 1912 with the "Hydro Circus."

5. The Queenston-Chippawa Project   69
Comparable in scope to the Panama Canal, Hydro's construction of the world's largest generating station fuelled Ontario's post-World War I industrial development.

6. The Radials Controversy   79
Three Royal Commissions in three years diminished Beck's influence and proved fatal to his dream of an intercity electric rail network.

7. Public Power in the United States   87
Ontario Hydro's success gave inspiration to the U.S. public power movement and New Deal President Franklin D. Roosevelt.

8. The Politics of Power 97

In pre-World War II Ontario, Hydro's struggles to keep up with rapidly growing demand for power were plagued by political interference and manufactured controversy.

9. The Natural Gas Betrayal 109

Federal Liberal and provincial Conservative decisions to give the young natural gas industry to private interests touched off an energy "civil war" and led to Hydro mistakes that still burden Ontario.

10. Nuclear Ontario 121

By the mid-1970s, it was clear to many that the frenzied push for nuclear power was an economic and environmental disaster in the making. Conservative and Liberal governments ignored the warnings.

11. The NDP Years: 1990-1995 139

The Rae government stopped Ontario Hydro's nuclear madness, changed its focus to energy efficiency, stabilized its escalating rates and reversed its growing debt.

12. The British Disaster 153

Privatization and deregulation in the United Kingdom's power sector were meant to fix its inefficiencies. The purported cure turned out much worse than the alleged disease.

13. The American Disaster – Part 1 163

Deregulation in the U.S. had its origins in global oil politics, expensive nuclear mistakes and federal government efforts to strip regulatory power from the states.

14. The American Disaster – Part 2 171

Things have gone so badly for deregulation in the U.S. that less than half of all states have opened their retail markets and most of the others wish they never had.

15. The Ontario Disaster – Part 1 187

The Harris Conservatives inherited a reinvigorated and financially healthier Ontario Hydro, but continued to insist that regulated public power was a bad idea.

16. The Ontario Disaster – Part 2   199

Despite years of warnings that deregulation would lead to power shortages and higher prices, the Eves government plunged ahead with what became the world's shortest experiment with retail electricity competition.

17. The Inevitable Failure of Deregulation   213

The unique nature of electricity, combined with basic principles of prudent investment, makes it impossible for deregulated markets to sustain lower and more stable average power prices than regulated markets.

18. The Retreat from Privatization   229

Recent efforts at electric utility privatization in Canada and the United States have fizzled, mainly because of public opposition. The public has been proven right.

19. A 21st Century Public Power System   243

Ontario's future power system must serve our province's economic and environmental needs. Public power is the only way we can ensure democratic control over our energy future.

20. Our Gift to the Future   259

The vision of our public power pioneers helped make Ontario one of the world's most prosperous regions in the 20th century. It is our turn to do the same favour for generations to come.

Notes   263

Sources   284

Index   287

Foreword

Why write a book about public power? The obvious reason: this is an important public issue, one that is fundamental to our economy, our environment, and to any semblance of social order in our daily lives. When our hydro bills go through the roof and we hear public announcements about the risk of power outages, these fundamentals become very real for all of us.

This is also an issue on which the party I lead, the Ontario New Democratic Party, has taken a very clear, consistent and passionate stand: public power must be preserved, even as it must change. And because this is such a complex issue, I needed a book to explain why.

There are also personal historical reasons for this book. In the early 1950s, my father worked as a line construction contractor for Ontario Hydro as it completed the decades-long task of bringing "the Hydro" to farms and rural communities everywhere in this vast province. I followed in my father's footsteps, for a while at least. As a university student in the 1970s, I worked for Hydro Construction building new transmission towers that hopped from island to island as they crossed Rainy Lake in Northwestern Ontario. To this day, when I drive by those huge towers I say to myself: "You helped build all of this." I know my father feels the same whenever he sees stretches of the hundreds of kilometres of hydro lines he helped to build as much as fifty years ago. Neither of us wants to see our work, and that of so many others like us, sold off and used for private profit.

I also come from a community where some of Canada's important legal history on power contracts has been made. My hometown of Fort Frances has always received its electricity from hydroelectric power dams owned by the resident paper mill. Twice in the twentieth century, the paper company went to court to get out of the "1905 Power Agreement" or at least reduce its obligations to provide power at cost under the agreement. In 1925, the issues were decided on appeal by the Divisional Court (*Town of Fort Frances v. Fort Frances Pulp and Paper Company* (1925), 28 O.W.N. 402), and in 1983, the same issues were decided at the Supreme Court of Canada (*Fort Frances (Town) v. Boise Cascade Canada Ltd.*, [1983] 1 S.C.R. 171).

As a high-school student, I spent my summers working in tourist camps and resorts far off the beaten track. Getting to some of them meant travelling 30 to 40 kilometres by boat across the lake. For one job, I had to fly more than 60 kilometres by float plane, an exhilarating experience, especially for a teenager. None of these tourist resorts were on the Hydro line. For power, they depended on noisy, expensive and not always reliable diesel generators. Guess who usually had to get up at 5 a.m. to crank up the diesel so the guests would have power in the morning? And who got to shut it down in the wee hours, after everyone else went to bed? Not quite as exhilarating. I learned every early morning and every late night how important it is to have power and how different life is without it! Since then, I have never taken reliability for granted. It has always been my greatest Hydro-related concern, personally and politically.

Finally, as a law student who became attracted to labour law, I spent the better part of a year on a research paper analyzing labour relations and collective bargaining at Ontario Hydro. I interviewed former leaders of CUPE Local 1000—the Power Workers' Union—and a former labour relations officer from management as well. The history of collective bargaining and labour relations at Ontario Hydro would be the basis for a fascinating book in itself.

These strong personal connections I have to Hydro-related issues helped supply me with perspective and motivation as I worked on this book. But the biggest motivators of all were my children and the future they will inherit, the future *all* our children will inherit. What happens to our economy, our society and our environment is inextricably interwoven with what happens to our electricity system—our Hydro. If we want our province and our world to be a better place for our children, and their children, we must pay close attention to how we can best provide, now and in the future, this most essential of public services.

Introduction

One hundred years later, public vs. private power is once again a dominant economic and political issue. Will the outcome be different this time?

Government should not be involved in the electricity business. Everyone knows the public sector is not as efficient as the private sector, nor as innovative. Moreover, political interference, such as regulating electricity rates, distorts the economics of the power industry. Market forces may be ignored, but they cannot be defeated, and we will ultimately regret even trying. While there may be some role in the electricity system for government, such as ensuring a level playing field in an open marketplace, public ownership and operation of system assets is a bad idea that inevitably leads to unsustainable public debt.

These are the now-familiar claims that you and I have been hearing for years from those advocating the privatization and deregulation of Ontario's publicly owned power system. Ironically, they are also precisely the same claims made one hundred years ago, at the dawn of the electricity age, by private power interests opposed to the formation of a "people's power company" in Ontario. Such a scheme would be the ruination of the province, they argued, and then they did their utmost to make this prediction come true.

As we know, corporate power interests failed back then to defeat public power. But have they finally won, a century later? After all, Ontario now has a deregulated wholesale power marketplace and the current government is scheming to sell off billions of dollars worth of publicly owned system assets. The retail deregulation scheme put in place on May 1, 2002, has been a dismal failure. Still, it has not been killed, only put on hold until after the next election. The new corporate power interests and their political representatives endlessly repeat the mantra that the privatization and deregulation of our hydroelectric system cannot be turned back. Ad nauseum they intone: "We have no choice but to privatize and deregulate."

The message of this book is that the opponents of public power are as wrong today as they were a hundred years ago. In those days, affordable, reliable electrical energy was essential to the economic development of the province and the social progress that would follow in its wake. Private power developers, though entrepreneurial and industrious, gave no sign that they would be guided by notions of the public good. That was government's mandate; theirs was to make money for their shareholders.

The people of Ontario—through their elected representatives and sometimes even through direct referendum—decided that public need must trump private greed. History proved them right. The founding of the Hydro-Electric Power Commission in 1906 quickly led to dramatically lower energy costs in the province and the rapid expansion of what would soon become one of the world's largest, lowest-cost and most reliable power systems. While not without its problems, public power is arguably the single most important factor in Ontario's remarkable and sustained economic growth throughout the twentieth century.

Today, affordable, reliable electricity continues to be a cornerstone of Ontario's economic health. No one doubts this. There is only disagreement on whether the public or private sector can better fulfill this need. The energy-related challenges we now face, however, go beyond power prices and availability. An environmental stew, that seemed to cook in slow motion for years, is now boiling over. Global climate change and local human health problems are warning us loudly that the effects of pollution are reaching critical mass. The power industry is a large part of the problem and must be an equally large part of any solution. At the same time, a technological revolution is underway in both the demand and the supply sides of the electricity sector. Will Ontario be a world leader in this revolution, as it was in the movement that electrified much of the world in the last century? Or will we become merely the buyers and not the inventors and sellers of new technologies that produce more environmentally benign energy and use it more efficiently?

The opponents of public power today say that the private sector is the most effective agent of change in meeting all these challenges, old and new, and that deregulation is essential to unleashing the benefits that will flow from individual enrichment initiatives. Regrettably, they add, there may be some transitional hiccups (California, Alberta, Montana, Enron, heat waves, nuclear plant shutdowns, etc.) as we move into this brave new world of market discipline. But don't worry, everything will sort itself out and all of us will ultimately be better off.

History is already starting to prove the opponents of public power and regulation wrong. North American power prices are unstable. System reliability, which we haven't worried about for decades, is now a front-burner issue. Fossil fuel emissions are steadily increasing. More people—and other species with whom we share this planet—are dying of pollution-related causes. Who knows how many more will die, or have their lives tragically altered, because of climate change caused or exacerbated by that same pollution? And sadly, Ontario is now falling behind in energy technology research and development; our former status as a global centre of power industry innovation has dramatically shrunk in recent years. There is no longer any publicly owned research facility like the former Ontario Hydro Technologies, now downsized and privatized. Great energy-related ideas that may take years to commercialize, are now, in Ontario, almost exclusively the domain of large corporate interests that couldn't care less whether their product development and manufacturing happens here in Ontario or somewhere else.

Once again, at the turn of a new century, the people of Ontario have the power to proclaim, through the ballot box, that private power interests cannot be allowed to diminish or defeat the public good. The corporate forces behind deregulation and privatization in Ontario, though they have won some recent victories, are not as confident of ultimate success as they once were. Public opinion is overwhelmingly opposed to their agenda. But the Conservative government's humiliating suspension of small customer deregulation in November 2002 and its cancellation of the Hydro One sale in January 2003 are—and everyone knows this—only strategic retreats to get past the next election. Should they retain power, their plans to completely dismantle this great public heritage and turn it over to corporate interests would proceed very quickly. The Liberals, with their grand boasts to Bay Street that they have always supported an open power market, would be no better. Only voters can put a permanent end to this tragic experiment, a failure everywhere it has been tried, and prevent privatization and deregulation from becoming entrenched and irreversible here in Ontario. We have the power to regain control over how our province's electricity system evolves. We have the right, as a society, to use our hydro system to pursue our common objectives: stable energy prices, great reliability, economic growth, regional development, technological excellence, environmental protection and, perhaps most important, public accountability for all of these things.

To reassert public control over the electricity industry would not be turning back the clock. It would be taking charge of our energy future, just as prior generations did.

There will be important roles for the private sector to play in the future of our electricity system, as there always have been. The private sector will continue to supply the turbines, the cables, the switches, the computers, the transformers, the steel girders, the insulators, the vehicles, the tools, the fuel and the thousands of other materials and technologies needed to run a power system as large and complex as ours. It will continue, as well, to provide Ontario utilities with an enormous range of essential expertise: technical, engineering, administrative, financial, legal, marketing and more. Private investors attracted to the security and stability of government-backed bonds will also be a primary source of future capital financing needed to upgrade and, when necessary, expand our system.

Imagine a world without wires. New technologies, from fuel cells to solar cells, will put the capacity to generate power into every building, an exciting development that promises many positive environmental and economic benefits.

It's technologically possible but still economically distant. Central power plants will continue to be the main source of electricity for many decades to come. In this context, there may be practical reasons to allow some degree of privately owned generation that can contribute to our province's environmental well-being within a regulated public power system.

I am not ideologically opposed to private power any more than I am opposed to private restaurants, clothing stores or car dealerships. My opposition to private power, especially if deregulated, is experience-based. Ontario's own unfortunate experience only adds to the body of evidence against deregulation and privatization. But that doesn't mean that governments must own every last nut and bolt of Ontario's electrical system. For example, I am very excited by the prospect of more private green energy projects, such as the cooperatively owned wind turbine at Exhibition Place in Toronto. I am also a huge fan of energy efficiency retrofitting, which depends on the private sector to be delivered. In my view, the evolution of our province's public power system should be open to new organizational concepts, new technologies and new approaches to addressing our perennial concerns about reliability, affordability and environmental protection.

All progress involves change. Like all institutions, our public power system must evolve. Its role is to serve our society's needs. When they change, so must it. And in fact, throughout much of its history, Ontario Hydro *has* responded to technological, economic, environmental and democratic forces. Maybe not always as quickly as it should have, sometimes not at all and sometimes poorly. The great weight of evidence, however, shows that public power in general, here and elsewhere, has done a better job than private power of addressing those needs, changes and forces. Where it has fallen short, the problems can always be traced back to a lack of government leadership and vision.

Ontario's electricity system is in trouble, the victim of free-market ideology, compounded by years of poor or non-existent political leadership. The Conservative government of today has no real vision of the future of our system, just a vague notion that everything will eventually work out all right if we simply let impersonal, uncaring market forces do their job. The Liberals appear to change their minds with every poll. New Democrats, however, have been consistently opposed to deregulation and privatization. We do not want to give up our right as a society to take the energy path that we democratically decide to be best for our future. We need regulation in order to shape our system so that it best serves Ontario's needs, not the profits of private shareholders who may not even live here and therefore won't care what happens to Ontario. We in the NDP want to regain control over our energy future and renew our once-great public power system. We know that most Ontarians feel the same way.

Before we set about building our twenty-first century public power system, however, we would be wise to try to understand how we got to where we are today. While much has changed since the early years of Ontario's electricity industry, much has not. We ignore the lessons of history at our peril and we impoverish our culture by forgetting or diminishing the great achievements of those who came before us.

1

The White Coal of Niagara

At the dawn of the electricity age, private interests came very close to long-term monopoly control over Niagara's immense hydroelectric potential.

The creation of the world's first publicly owned long-distance power system is an inspiring story of the power of democracy and political action. Indeed, it is one of the most compelling political stories in Canadian history. From its first stirrings in the early 1890s to that historic day in 1910 when non-profit electricity generated at Niagara Falls first lit an Ontario city, the movement for public power transcended all social, economic and class differences. An unprecedented combination of farmers, workers, business owners, politicians, the rich and poor, the old and young, no matter where in the province they lived, were united in one purpose: wresting control of Ontario's infant electricity system from private power interests whose greed and lack of foresight were greatly hampering economic development. That it took as long as it did to accomplish this goal is a parallel story of the power of money to influence public policy, especially in a government without principles. Time seems not to have diminished money's power, but our history proves that it can be overcome. Ontarians did it one hundred years ago and we can do it again today, in exactly the same way.

The first public policy steps that ultimately led to Ontario's public power system had nothing to do with electricity, but everything to do with Niagara Falls. By the 1870s, Niagara Falls was already known as the Honeymoon Capital and was one of the most popular tourist sites on the continent. As the tourists flooded in, so did the vendors, the developers and the hucksters, honest and otherwise, giving the area a chaotic and lurid atmosphere that became seedier with each passing year. Many civic leaders, on both sides of the Falls, called for public control over development of the areas bordering the Niagara River. In 1878, Canada's Governor General, the Irish-born Lord Dufferin, advocated the creation of an international public park at Niagara Falls in a speech that was as widely hailed in New York as it was in Ontario. The Dominion Government heeded his call and in 1880 passed the "Act Respecting Niagara Falls and the Adjacent Territory," giving Ottawa the right to acquire lands for park purposes. It was the provincial government, however, that took the financial initiative. In 1885, the Ontario legislature passed "An Act for the Preservation of the Natural Scenery about Niagara Falls." A commission was established, chaired by Sir Casimir Gzowski, with John Grant Macdonald, and John Woodburn Langmuir as members. In 1887, the Commission named its project the Queen Victoria Niagara Falls Park and issued bonds that were backed by the Province, a money-raising method later used to fund the development of Hydro. The Commission used the proceeds of the bond sale to acquire the lands bordering the Niagara river. It also began to clean up and regulate the tourist industry. Wax museums, fudge shops, boarding houses, souvenir stands and carnival booths were fine, but not on the banks of the river, obscuring the splendour of the Falls. It was in this government-driven initiative to preserve natural beauty for future generations that the origins of public ownership and control of Ontario's share of Niagara's power are to be found.

In 1871, when Belgian inventor Zénobe Gramme began to manufacture the first reliable "dynamo" for power generation, not a single watt of electricity was being produced for sale in North America. Ten years later, the Brush Electric Light Company installed one of the first hydroelectric generators in the world on the U.S. side of Niagara Falls. Charles Brush had invented reliable arc lighting just before Thomas Edison commercialized the incandescent bulb. Like Edison, whose first central power station was installed in New York City in 1882, Brush developed a generating system to light his lamps and Brush street lighting systems were quickly taken up in several U.S. cities. The tiny Niagara plant operated 16 bright arc lights, adding to the area's tourist attractions. Prior to the development of artificial lighting, electricity had been used mainly to operate telegraphs and the recently invented telephone. Brush and Edison changed all that, although their direct current (DC) revolution would not truly take off until the later inventions of the brilliant Serbian-American Nikola Tesla made alternating current (AC), electrical transformers and high-voltage transmission possible. In popular history,

In 1903, the privately owned Electrical Development Company of Ontario began work on its generating station at Niagara Falls to supply the Syndicate's Toronto Electric Light Company. As is apparent in this picture, construction safety standards were rather low back then; fatal accidents, including drownings, were not uncommon. The plant was eventually purchased by Ontario Hydro in the early 1920s. *Hydro One Archives*

Edison is given much of the credit for lighting the world, but Tesla's breakthroughs made modern power systems possible.

Electricity generation came to Canada at about the same time as in the U.S. In 1878, the Canadian-owned American Electric and Illuminating Company built a generating plant in the shopping district of downtown Montreal. This was the first report of electricity for sale in Canada. In 1881, the Ottawa Electric Light Company operated three small generators driven by a water wheel to power arc lights in a handful of mills. By 1884, Peterborough, Hamilton and Toronto had installed electric street lighting, the first Canadian cities to do so. In Toronto, the two small generators that powered arc lights on small stretches of Yonge, King and Queen Streets at night, were soon turned to operating labour-saving devices for the home during the day, an innovation that presaged the true scope of the age of electricity. Toronto was also the first city in the world to develop an electric railway. In 1883, a one-mile line in the western part of the city took passengers to the Toronto

Exhibition Grounds. Around this time as well, the Parliament Buildings in Ottawa changed from gas to incandescent lamps, and arc lamps lit many of the surrounding streets (except on moonlit nights, when they were deemed unnecessary). MPs returned to their home ridings extolling the wonders of electricity and urging their constituents to embrace its benefits.

As with many socially transforming technologies, however, reaction to electricity was mixed, with some calling it the invention of the devil. One Toronto minister fulminated from his pulpit that the electric motor was an instrument of evil since it would free girls from honest toil and allow them to wander about the streets, inevitably falling prey to the wiles of Satan. Such warnings were ignored. By 1890, most Ontario towns of over 3,000 people had some type of electric lighting station in operation. Each community required its own generator because the technology to transmit electricity over long distances did not yet exist. The power fed by early generating stations into the local electricity distribution network quickly dissipated with distance, in much the same way that a flashlight beam loses brightness the further it travels from its source. Most local power plants were coal-fired. Coal was burned to heat water to create steam to turn turbines. There were also a number of small hydroelectric (or hydraulic, either term is correct) stations, like the one in Ottawa, in communities situated on rivers. In these stations, falling water turned the turbines. The volume and speed of falling water determines how large a turbine it can turn and, thus, how much electricity can be generated. Since the narrow Niagara River was the main funnel for Lake Erie water as it fell some 100 metres into Lake Ontario, the potential energy in that water was very great. Without a long-distance transmission system, however, that energy—the "white coal" of Niagara—would remain out of reach for all but a handful of nearby Ontario homes and businesses.

Scientists and engineers doggedly worked at solving the transmission problem. By 1896, enough progress had been made so that the Niagara Falls Power Company, which had been established on the U.S. side a few years earlier, began sending power from the Falls to the booming industrial city of Buffalo, 20 miles away. Word of this development spread fast in Ontario, where municipalities with coal-fired systems began to look yearningly towards Niagara. "If the Americans can do it, why can't we?" This question took on new urgency in 1897, when a Pennsylvania coal miners' strike led to skyrocketing fuel prices and power shortages in a province with no coal of its own. With electricity costs high on the public agenda, attention was increasingly directed to the activities of private power developers looking to lock up generation rights on the Canadian side of the Niagara River.

Almost from its inception, the Queen Victoria Niagara Falls Park Commission had been trying to sell hydraulic generation rights to private developers. There is irony in this first-ever example of electricity sector privatization: here is a public body charged with protecting public assets trying hard to sell them off. The irony turns to horror, however, when one learns that in 1892 the Commission gave a twenty-year contract to the American-owned Canadian Niagara Power Company for the exclusive right to use Niagara water on the Canadian side to generate electricity. The contract contained an option to renew for four further twenty-year periods, effectively giving the U.S. company a one hundred year monopoly. There can be no doubt that the history of Ontario would have been much different had this "deal of the century" gone through. Fortunately for Ontario, the company ran into financial problems and failed to start construction by the contractually promised date of May 1897. The company explained that it had not been able to solve the transmission problem, but this was a transparent lie. Its U.S. parent was the Niagara Falls Power Company that had built the line to Buffalo the preceding year. The company appealed the contract expiry and, in a controversial decision, the High Court of Justice for Ontario granted an extension. The company, ruled the court, could retain its monopoly if it could deliver 10,000 horsepower by November 1899. Once again, the company failed to meet its deadline and the Commission, directed by the Ontario Legislature, took back the monopoly rights it had foolishly given away seven years earlier for a mere $25,000 a year.

In defence of the Commission it should be pointed out that the idea of a publicly owned electricity system was a very dim star on the economic horizon at the time. Moreover, the Commission did make some effort to protect Ontario's interests. The contract stipulated that Canadian Niagara Power had to make at least half of the power it generated available to Canada, at whatever price the company charged Americans. While some commentators at that time criticized even this modest regulatory measure as unwarranted government interference in industry, many newspapers, civic leaders and business interests castigated the Commission for its 1892 surrender of Niagara's boundless natural wealth to American interests. From that year forward, right up to the present, electricity has been a prominent political issue in Ontario.

Having escaped the awful prospect of an unregulated private sector monopoly, the Commission nonetheless persisted in its efforts to privatize Niagara's power. In 1899, it offered the hapless Canadian Niagara Power a 50-year, non-exclusive concession to a significant portion of

Canada's water rights at Niagara. Predictably, the company missed its new contractual deadline of July 1903, which the Commission then duly extended to January 1905. One wonders why a consistently non-performing company was kept in the game by the Commission for more than ten years. Let us give the commissioners the benefit of the doubt and presume that they had honest reasons for doing so. If a lack of alternative developers was one of those reasons, this soon began to change. In 1900, another U.S. developer, the Ontario Power Company, won a similar 50-year license from the Commission to use water from the Welland and Niagara rivers to generate electricity. Less than three years later, the first Canadian power developers arrived on the scene. A syndicate headed by William Mackenzie, a key figure in the building of the Canadian Pacific and other railroads, obtained a license to exploit Niagara for power on the same terms as the Americans. Also in the syndicate was Henry Mill Pellatt, who eventually went bankrupt after building Casa Loma, then the largest private residence in North America and now one of Toronto's main tourist attractions. The Mackenzie Syndicate already controlled the Toronto Street Railway, which had a 30-year contract to operate the city's streetcar system. It also owned the Toronto Electric Light Company, which held sole rights to install all street lighting in the city. Not satisfied with these monopolies, the Syndicate wanted to control *all* the power that would *ever* be brought into Toronto and was intent on building its own transmission line from Niagara.

If Ontario's nascent electricity industry in 1902 had continued to evolve as it had during the previous decade, the prospects for public power would almost certainly have faded into oblivion. Instead, a remarkable political coalition emerged that defeated a government, stopped the privatization of public assets and set the stage for the development of a public power system that was soon widely imitated throughout North America.

2

The Triumph of Democracy

Failing to respond to growing popular demand for public power, the Ross Liberal government was crushed at the polls by the Whitney Conservatives.

By the mid-1890s, it was clear that electricity was destined to be the single most important ingredient of economic progress. In the emerging industrial region of southwestern Ontario, though, this new source of energy was in increasingly short supply. Demand for electricity had been growing exponentially, but private investment in additional generation and distribution capacity had not been keeping pace, especially outside the Toronto-to-Niagara population belt that hugged the western shore of Lake Ontario. The inevitable consequence, in the absence of regulation, was high power prices. The power industry's freedom to charge what the market would bear was becoming a great burden to the rest of Ontario's economy. This problem would resurface a century later, in 2002, when power producers were once again allowed to set their own prices.

Ontario desperately needed power. There were more than a thousand industrial plants in the province. Most of them were operated by steam engines that moved the belts and pulleys and pistons of the production process. Steam engines were less energy efficient and far less flexible than electric motors, but without a reliable supply of affordable electricity, many factory owners judged it too risky to make this technological leap. Moreover, newer industries that were already based on motors needed to locate where power was cheapest and most reliable. Many feared that if the electrification of Ontario industry did not keep up with its American counterparts, and if power continued to be expensive, the province would soon become an economic backwater. These fears were well-founded. Private power developers at the turn of the last century were seldom motivated to invest in more and better service to smaller and more geographically scattered population centres. Returns on capital were much more attractive in the larger cities. This was especially true for the power that would eventually come from Niagara. With transmission technology steadily improving, it would soon be possible to generate energy at the Falls and deliver it hundreds of miles away. By 1902, the Mackenzie Syndicate was willing to fund the construction of such a transmission line from its planned generating station at Niagara, but not to areas where power sales were still too low to be lucrative. The line would be built mainly to supply the Toronto region, whose electric utility and streetcar service, it will be recalled, were also owned by Syndicate members. Then, as now, corporations had little interest in investments that might take years to bear fruit when better returns were immediately available.

Believing it the single most important factor in their community's future economic development, civic leaders throughout southern Ontario fixated on the issue of electricity supply. So did local business associations, both industrial and commercial, who knew that the steam era was coming to a close. So did the unions of the time, seeing in electricity the promise of an end to the hazardous drudgery of dark mills and the age-old oppressiveness of cold, dimly lit plants, shops and homes. So did the vigorous progressive movements of the day, including early farm associations, which saw in electricity a powerful force for what we would now call economic democracy. Many also foresaw its potential for greater public health and safety. Indeed, safety was often uppermost in the minds of those who preached the benefits of electricity. In a world lit only by fire, out of control fires were common, as was their often tragic aftermath. A century ago, electric lights were nothing less than miraculous. While giving off many times more light than gas or paraffin, they dramatically reduced the risk of fire and thereby saved many lives and much property. Brightly lit streets made people safer at night. More frequented streets added a new dimension to community life. All of electricity's benefits, however, were in the hands of a small number of private companies that used their monopolies to reap profits that were large even by the standards of those times, when the big corporate "trusts" in mining, steel, banking, railways, meat-packing and other sectors, controlled most key North American industries.

In Toronto, the idea of a municipally run utility, to compete with Henry Pellatt's Toronto Electric Light Company, surfaced in the final years of the nineteenth century. Its chief advocate was Toronto Alderman F.S.

Spence, who urged Toronto City Council in 1899 to apply to the provincial Liberal government of George Ross for the right to build a city-owned generating station. Based on engineering studies, councillors were confident that they could provide power at much lower cost than Toronto Electric Light. They were supported in their cause by two well-known Toronto businessmen, P.W. Ellis and W.K. McNaught, who had recently visited Switzerland, then the world's most advanced country in industrial electrification due to its inexpensive hydroelectric power. On their return, they urged the city to construct its own generating station at the Falls, its own transmission lines to the city, and even its own distribution system.

Queen's Park was a short walk from Toronto City Hall, but they may as well have been an ocean apart on this subject.

In 1899, Toronto Alderman F.S. Spence urged development of a municipally-owned electric utility to compete with the Syndicate's expensive power. The Liberal government repeatedly turned down the city's request for legislation allowing a non-profit utility. It later came out that Premier Ross was an investor in the Syndicate's private power company. *Hydro One Archives*

Premier Ross was not willing to allow public competition with existing private enterprise and turned down Toronto's request. This was consistent with Liberal policy of the time. The Municipal Act already barred municipalities from competing with incumbent private water companies. In 1899, the Ross government amended the Act to explicitly forbid municipalities from competing against existing electric utilities as well, thereby entrenching the private monopolies. To be fair, the Liberals did not, strictly speaking, outlaw public power. They left a loophole, but it was a very expensive one for municipalities seeking to provide local homes and businesses with affordable, dependable power. Under the Act, a municipality could apply to buy out an existing private utility, at a price to be arbitrated. This greatly raised the cost of entry into the power marketplace and squashed the hopes of most municipalities for the affordable power they so desperately needed.

For years, Daniel B. Detweiler, left, rode his bicycle from town to town preaching the gospel of public power. In 1900, he teamed up with Elias W.B. Snider, right, a successful flour miller and politician, to lead Ontario's municipal power movement. *Hydro One Archives*

The private utility lobby, led by Pellatt, had the Liberal government in its pocket. Refusing to give up, the proponents of public power fought back.

As the new century dawned, Spence and others of like mind began organizing the municipal power movement, a common cause that lifted cooperation between Ontario municipalities to levels not seen before, and maybe not since. Two key figures in this cause were Elias W.B. Snider of St. Jacobs and Daniel B. Detweiler of Berlin (which changed its name to Kitchener in 1916). Snider was a successful businessman and politician, the first miller in North America to employ the revolutionary Hungarian rolling-mill process. He had also been a member of the Ontario legislature from 1881 to 1894. Detweiler never went to high school and began his working life as a shoe salesman, but quickly rose to a partnership in his company. Snider designed the system of municipal cooperation that fuelled the early years of the movement; he was a frequent speaker on this subject at business and political events. Detweiler was more of a grassroots evangelist. For years, he rode his bicycle throughout the Grand Valley, speaking tirelessly to all who would listen, of the day when people would have electricity to light their homes and operate their factories. In 1900, the two began corresponding and soon became the dynamic duo of public power in southwestern Ontario, with Spence leading the forces in Toronto.

At the same time, progressive forces in other provinces were also organizing for public power. In 1901, the year of Queen Victoria's death and Marconi's wireless transmission from England to Newfoundland, representatives of Canadian municipalities met in Toronto to discuss how to speed up the development of electricity distribution and control its price. Calling themselves the Union of Canadian Manufacturers, they elected O. A. Howland, Mayor of Toronto, as President and undertook an aggressive lobbying campaign in Ottawa, Queen's Park and other provincial capitals. (This was also the year that 63-year old Annie Taylor became the first person to go over Niagara Falls in a barrel. She survived, but many of her imitators didn't and such stunts were later outlawed.)

Nearly all Ontario newspapers, regardless of political affiliation, favoured public power, but the Syndicate still had the provincial Liberals in its pocket. In 1902, when Toronto Council asked the legislature a third time for authority to generate and distribute electricity, the answer again was no. Even the enthusiastic support of the Toronto branch of the Canadian Manufacturers Association for a publicly owned municipal utility made no impression on the government.

Reading the mood of the people, the opposition Conservatives latched onto public power as their chief political issue. In 1902, just before a provincial election, they sponsored the following resolution in the legislature:

> In all future agreements made between the Commissioners of the Queen Victoria Niagara Falls Park and any other person or persons, power shall be reserved to the Provincial Government to at any time put a stop to the transmission of electricity and pneumatic power beyond the Canadian boundary; and that in the opinion of this House, the water of the Niagara River and its tributaries, as well as the waters of other streams, where necessary, should at the earliest moment and subject to existing agreement, be utilized directly by the Provincial Government in order that the latter may generate and develop electricity and pneumatic power for the purpose of light, heat and power, and furnish same to municipalities in this Province at cost.

Thirty Conservative members voted for this nationalistic "Power at Cost" resolution. It lost when 41 Liberals voted against it.

Another Pennsylvania miners' strike that same year caused public interest in Niagara power to surge yet again as the situation for Ontario's manufacturers became disastrous. Coal that had cost $3.50 a ton was now hard to get at three times that price. For a while, Toronto imported coal from Wales at $10 a ton, but this was clearly not sustainable. Niagara had to be tapped. Public power had become a major public issue.

In the provincial election of 1902, the once-large Liberal majority of Premier Ross was reduced to a majority of one. There were several issues in the election, but most newspapers attributed the results to the Liberals' persistent opposition to public power. Elected to the legislature for the first time was Adam Beck, who had twice been mayor of London, the largest city in southwestern Ontario.

Sensing political advantage, the leaders of the municipal power movement quickly convened an all-day meeting on June 9, 1902, at Berlin's Walper House, which still stands at the corner of King and Queen Streets in downtown Kitchener. In attendance were 25 civic and business leaders, mostly from cities that had reason to worry about being disinherited from the riches that would soon flow from Niagara: Preston, Hespeler, Waterloo, Berlin, St. Jacobs, Guelph, Galt, Bridgeport and Brantford. The keynote speaker was Toronto Alderman Spence, who laid out in his address all the essential elements of a public power company that would later be adopted by the next government. What was required, Spence said, was:

> ...a government commission which would have power to arrange for the transmission of electricity to the various municipalities desiring it; this commission to issue its own bonds in payment of transmission lines, which bonds would be covered by bonds of the municipalities interested. Under the scheme, the government, through a commission, would undertake the transmission to the municipalities desiring power, the latter guaranteeing by their bonds the cost, and selling in turn to all manufacturers at an even rate, preventing in this way the power from falling into the hands of any monopoly, and in this way securing to the industries of this Province advantages of cheap electrical energy.

An action committee led by Snider and Detweiler was formed to organize intensive lobbying of the Legislature by municipal and industrial leaders. Their goal was to make real the vision Spence had articulated, a municipal electric cooperative that would build a transmission system linking every industrial city in southern Ontario to the power of

Niagara. They needed legislation to make this possible and they had to move fast. The Mackenzie Syndicate was attempting to gain the remaining water rights franchise at Niagara.

A personal meeting with Premier Ross in October only confirmed the Liberals' aversion to public power. He dismissed the worries of the Snider delegation that they would be hostages to price-gouging private companies. The Niagara Falls Park Commission had said that the planned private development of more than 400,000 horsepower would provide plenty of surplus power for well into the future, 20 years or more. Moreover, with three companies in the business, even though two of them were American, there would be sufficient competition to keep prices low. It was therefore no surprise to the Snider committee when, in January 1903, the government granted an irrevocable franchise to the Syndicate to generate 125,000 hp—at an annual payment of only $15,000 for water rights! The idea of a municipal electrical cooperative had been rejected once again by the Liberal government.

Adam Beck, (1857-1925). Beck, whose paternal grandparents emigrated from Baden (now in Germany) in 1829, was a self-made business success and two-term mayor of London before his first election to the Legislature in 1902, around the time this picture was taken. *Hydro One Archives*

Public power advocates were outraged. The Syndicate was painted as a band of robber barons who would put the people of Ontario in perpetual economic bondage. Why, it was loudly asked, should a handful of avaricious men reap huge profits from a natural resource that rightly belonged to the people? How long would the Liberals continue to deny municipalities a share of the water power that had been given away to "predatory and parasitical promoters" who provided bad service at high cost? The answer to these questions was money. While he was Premier, Ross was also president of the Manufacturers' Life

Insurance Company. Fellow officers and board members included the principals of the Syndicate's Electrical Development Company. Three years later, it would come out that Manufacturers' Life had invested in the Electrical Development Company. Ross had rejected the public power movement to enrich himself and his friends. No other interpretation of the facts makes sense. Government corruption emanating from the Premier's office attempted to kill the prospect of public power once and for all. There were no more water rights left at Niagara.

Fortunately, this attempt to thwart public power failed. As the protests gathered steam, backbenchers from cities with strong local public power movements pressed Ross to compromise. The movement had become an irresistible political force, the government was no longer an immovable object. Its one-seat majority hung by a thread. Among the harshest critics of the Liberals was the Toronto Branch of the Canadian Manufacturers Association. After the Liberal government turned down Toronto's request for its own utility a fourth time, the Association wrote the Premier in 1905 that it feared the private utilities would not charge fair prices. The Association forcefully declared its support for a municipally owned utility and pledged to work with the city to do whatever would "guarantee electric power and cheap light to the citizens of Toronto for all time to come at the actual cost of the same, or at a fixed percentage of profit upon the cash expended." Coming from an organization noted for its political and economic conservatism, this was a remarkable pronouncement.

Ross had to bend, so he wrote a letter to the Niagara Park Commission asking for an estimate of the cost of transporting power "within a practicable distance" of the Falls. This request was politically presented as a possible, but not guaranteed, precursor to a public power scheme. Clearly it was meant to buy time for the Syndicate, but its main effect was to stoke the optimism of the movement's leaders, who redoubled their efforts to amass public support. The Snider-Detweiler Committee held yet another meeting in Berlin, this time in the YMCA hall. Political and business representatives from ninety different Ontario communities were present. Among them for the first time was Adam Beck, whose attitude to public ownership had not been clear prior to this meeting. After this event, however, everybody knew where Beck stood.

The meeting was called to hear the results of the Snider-Detweiler Report, one of the pinnacles of the struggle for public ownership. The document mapped out a municipal transmission cooperative that would be able to deliver power at less than half the cost then being charged by the private utilities. The report was enthusiastically approved, but with one amendment advanced by London Mayor Beck and Toronto Mayor Urquhart, a Liberal lawyer and long-time advocate of public ownership. Snider and Detweiler had not suggested provincial government involvement in the cooperative, only legislation that would enable municipalities to do it for themselves. Beck and Urquhart's amendment, however, called for the "building and operating as a Government work, a line for the transmission of electricity from Niagara Falls to the towns and cities; and that the municipalities here represented call upon their representatives in the Legislative Assembly of Ontario to urge upon the Government to carry out this resolution."

Typical of the editorial comment on this resolution was the Toronto weekly, *Saturday Night*, which wrote that the Berlin convention "will inspire with new hope and purpose all advocates of industrial and political progress, who believe that the many were not designed to be forever bled and bullied by the few."

James Pliny Whitney (1843-1914). "The water power of Niagara should be as free as the air." Born in Morrisburg in eastern Ontario, Whitney was regarded as a remarkably honest politician for his time and was a fervent advocate of public power. He became the province's sixth premier in 1905 when the Conservatives he had led for eight years crushed the long-ruling Liberals. His three successive majorities were among the largest electoral victories in Ontario history. *Hydro One Archives*

Ten days later, a delegation led by Snider and Spence met Premier Ross at Queen's Park and put the case for public power to him again. If the government would not provide power at cost, let the municipalities do it, they said. The resource was owned by the province and administered by a government body. Where was the justification for denying its use to the public while private investors plundered it for profit? Ross again dismissed the delegation's concerns and repeated the claims of the private power interests that ruinous public debt would result from any public power initiatives. In the end, though, Ross gave in to at least some of the delegation's demands. He promised legislation that would enable the development of the municipal cooperative they so desired.

Months later, the promise was delivered, but it had so many strings attached that the idea would be unworkable for most municipalities. Under the Ross Power Act, as it became known, municipalities had the authority to buy, develop and sell electric power, but only if they could finance it themselves, without provincial government help. Since the Ontario Municipal Act severely limited the debt that municipalities could incur, few Ontario cities, let alone smaller towns, would be able to finance a public power system on their own.

For many, the Act was sham legislation designed only to give the appearance of support for public power while placing great obstacles in its way. It was, nevertheless, a step forward. The new law established the Ontario Power Commission to oversee the development of municipally owned power companies. Snider was appointed Chairman of the five-member body, which also included Adam Beck, Ross's nod to bipartisanship. The Commission conducted an exceedingly thorough study of the future power needs of industry and municipalities; the work took nearly three years to complete.

In the meantime, in 1904, the Syndicate's Electrical Development Company discovered that its proposed plant at Niagara would have access to much more water than had been granted by the Queen Victoria Niagara Falls Park Commission. The company asked the Commission for large additional generation rights, in accordance with an option spelled out in its original license. This option, however, carried the proviso that one half of the additional power would be reserved "for the use of municipalities, for the purpose of operating a municipal system of lights, heating or other public utilities." The company balked at this condition and tried to negotiate it away, much to the chagrin of Ross, who by now fully realized that he needed to be seen as supportive of public power. The Syndicate, however, wanted all the new power for itself. This time their greed proved to be their undoing.

Negotiations between Ross and the Syndicate were still underway in mid-December when a provincial election was called for January 25, 1905. Electricity was the dominant issue, especially in the cities. The Ross Power Act turned out to be too little, too late. In the largest-ever turnout in a provincial election up to that time, the Liberals were swept out of office after thirty-four years. The Conservatives won more than two-thirds of the Legislature's 98 seats, a smashing majority. Towards the end of the election campaign, the Syndicate finally accepted the 50 percent proviso because they saw the Liberals going down to certain defeat. Their manoeuvre, like that of the Liberals, was too late. The new Conservative government later cancelled the company's additional license on the grounds that it had not been approved by the Legislature.

The new Premier, James Pliny Whitney, had said in many campaign speeches, "The water power of Niagara should be as free as air." Once he took office in February 1905, he added these historic words:

> I say on behalf of the Government, that the water power all over the country should not in the future be made the sport and prey of capitalists and shall not be treated as anything else but a valuable asset of the people of Ontario, whose trustees this Government of the people are.

It is a tragedy that today's Conservatives have so completely spurned this noble declaration of the greatest achievement of their political forebears.

3

The People's Power Company

Led by the remarkable Adam Beck, public power forces dramatically defeated private power interests in two municipal plebiscites, in the courts, and at Queen's Park.

After the election, Whitney appointed Beck Minister without Portfolio, though all knew he was, in effect, Minister of Power. Beck had been re-elected by a large majority and was better known throughout the province than Whitney. He was already an independent political force, but gave his Premier no trouble. Beck was extraordinarily strong-willed, but seemed completely focused on electricity; he did not interfere in other areas of government concern. The two men agreed that the Ross Power Act was intended to ensure the failure of public power and promptly repealed it. In July, five months after the Conservatives took office, the Cabinet established the Hydro-Electric Power Commission of the Province of Ontario (HEPC) with Adam Beck as Chairman.

Beck's mandate was to survey the province's water resources that could be tapped for power, determine future demand and assess the capital cost of generation and transmission facilities needed to meet that demand, along with the rates that would have to be charged. It was a tall order that would normally take years to complete, but such a delay was politically unacceptable. Fortunately, the Snider Commission that had been appointed under the 1904 Ross Power Act had been diligently pursuing its own mandate for over a year and had done much of the research Beck needed. Much to his old ally's dismay, Beck used the spadework of the earlier Commission to bolster his own report,

On April 12, 1906, *The Globe* gave prominence to the report of the Beck Commission and its promise of "power at cost." *Hydro One Archives*

which was rushed into print just days before Snider's.

Not surprisingly, both commissions concluded that public power made far more economic sense than private power. Beck's report presented a simple argument that was difficult to refute:

> Competing [private sector] companies have to pay considerably higher interest rates on their bonded debt [than government], and in addition they have large issues of capital stock on which dividends have to be earned. Whether rates be fixed by the companies voluntarily or under Government regulation, regard must be paid to these conditions and the rates loaded accordingly.

This argument for public power is as valid today as it was back then. Electricity is one of the most capital-intensive industries. The price of money to finance facility construction and maintenance is the single largest component of the price of power. It is indisputable that government-backed utilities can borrow money at more favourable rates than can any private-sector company, no matter how large. Add in the requirement for profits and bloated executive salaries and private power becomes all that much more expensive. With these extra monetary burdens, private power operators, then and now, would have to be vastly more efficient than their public counterparts in order to compete. There is not a shred of hard evidence, however, that investor-owned utilities in North America are, in the aggregate, more efficiently run than publicly owned utilities. Those who make such claims typically rely on highly selective anecdotes that pander to the belief that the private sector is "naturally" more efficient. But if this were true, how then do we explain the loss of trillions of dollars from North American equity markets since early 2001? Though undeniably caused by private sector mismanagement, lack of foresight and often criminal dishonesty (which would arguably be much worse in the absence of government regulation), such economic disasters—and there are lots of them—are brazenly glossed over by the zealots of privatization and deregulation as "market forces at work." Tell those who lost their pensions in the Enron and WorldCom debacles that the private sector is more efficient. Government-backed Hydro bonds, as it turns out, would have been a much better investment.

The Beck Commission estimated that the HEPC would need four years to build a generating plant at Niagara, but recommended that this not be attempted immediately. The Syndicate's Electrical Development Company, among other privately owned generators, already had enough capacity to supply projected Ontario demand in the medium-term, provided acceptable rates could be negotiated. At that time, the Electrical Development Company was charging the Toronto Electric Light Company and the Toronto Street Railway $35 per metered horsepower per year. This was more than double the Snider Commission's $15.73 estimate of what non-profit power from Niagara would cost Toronto. Using slightly different assumptions, Beck's group pegged the cost at $17.

That public power was affirmed as cheaper than private power by both commissions was no surprise to municipalities, who were becoming impatient with such studies while their growing demand for power went unmet. In April 1906, representatives from 70 Ontario municipalities journeyed to Toronto to press their demand that the Conservatives' promise of public power be pursued with all dispatch. Fifteen hundred delegates jammed Old City Hall at Queen and Bay. To prolonged cheering and enthusiastic ovations for every speaker, they passed the following resolution:

That the municipalities now present and represented in the City

A 1,500-strong "power" deputation from 70 cities poured out of Toronto's Old City Hall on April 20, 1906 and marched up University Avenue to the legislature, chanting, "We want cheap power and we want it now!" *Hydro One Archives*

> Hall, Toronto, having an urban and rural population of over one million, respectfully urge upon the [Government] the necessity of safeguarding the people's interest by legislation enabling the Lieutenant-Governor-in-Council to appoint a permanent provincial commission with power to take, where considered by it advisable, the following action: The construction, purchase or expropriation of works for the generation, transmission and distribution of electrical power and light; to arrange with any existing development company or companies for power at a reasonable price, so as to be transmitted and sold by the Government to municipalities or others; also to vest in it the powers necessary to enable it to regulate the price at which electricity can be sold to all and every customer, whether municipal, corporate, or private.

It was no coincidence that this resolution closely resembled the legislation then being drawn up by Beck and his advisers. His promotion of the meeting and his closeness with the other leaders of the public power coalition were undisguised.

Four abreast, all 1,500 delegates marched up University Avenue and into the legislature, chanting: "We want cheap power and we want it now!" Whitney met with the leaders and assured them that his pledges would be honoured. The government would either generate the power municipalities needed or regulate the output and prices of the private companies. He asked for more time to study which approach would best serve the province. After all, the proposed cost of the transmission system alone was estimated at $15 million, a provincial expenditure of unprecedented size. Surely the municipal leaders saw the wisdom of great prudence before incurring such a responsibility?

The delegates left the Legislature reassured because Beck himself had previously told their leaders that legislation was imminent. The Toronto rally had been staged by the municipalities and orchestrated by Beck to drown out any lingering political opposition to public power and to impress upon Cabinet the necessity of moving faster. Both objectives were achieved.

On May 7, 1906, Beck brought in his Bill: *"An Act to Provide for the Transmission of Electric Power to the Municipalities."* It permanently established the Hydro Electric Power Commission of Ontario and gave it broad authority to pursue the goal of affordable power. Chastened by their rejection at the polls, the Liberals, who had done so much to frustrate and undermine public power, offered little opposition; most of them ended up voting for it. On May 14, the bill received Royal Assent and three weeks later, on June 7, Whitney appointed the three-member Commission: Adam Beck, John S. Hendrie, Mayor of Hamilton and a Cabinet minister, and the brilliant young engineer, Cecil B. Smith. Smith had been an investigator for the Snider Commission and, before that, Chief Engineer of the Canadian Niagara Power Company. Few people knew more about the physical realities and construction costs of power systems. Smith's resignation from the Commission in early 1907 and his subsequent defection from the cause of public power wounded Beck deeply. Smith's action was caused, in part, by Beck's often intransigent personality, but also by self-interest. Upon his resignation, Commissioner Smith immediately began earning large fees for consulting to companies anxious to ride, or stem, the tide of public power. Indeed, private power owners and developers had reason for hope. Beck had often said that Hydro would do business with them, provided Ontario customers would pay no more than American customers for the same power.

Abandoned for the moment by their former political sponsors, the Liberals, the Syndicate and its allies nevertheless continued their fight. They stepped up their accusations that public ownership could lead only to political patronage, graft, inefficiency, provincial bankruptcy and an era of anarchy akin to the collapse of the Roman Empire. Beck's promise to work with them, however, blunted much of their criticism. As the Toronto Mail and Empire noted in an editorial the day after the bill's passage:

> Those who described Mr. Beck's Power Bill as hostile to capital can have made but a very hasty examination of it. [It] does not stride over the claims of existing companies. It leaves the private

companies the opportunities to do all the power business on the market; and so long as they are satisfied with a reasonable profit and deal justly with all customers, the Hydro-Electric Power Commission will not interfere with them. It will leave them in possession of their franchise, their works and their power. So long as they adhere to fair rates it will have no cause to carry its regulative functions to the length of expropriating any power company's plant. But it is to be the judge as to what rates are fair.

It was true that the HEPC was given the power to regulate private power rates. But municipalities still held the key to whether the HEPC would succeed or fail. The Act allowed any municipal council to voluntarily enter into a power supply contract with the HEPC. Any such contract, however, would be provisional until it was approved by local ratepayers in a referendum. This highly democratic "opt-in" procedure set the stage for the next battle in the public power campaign.

A century ago, Ontario municipal elections were routinely held on New Year's Day, with the winners assuming office very shortly after the vote was counted. Local bylaw issues were often put to referendum vote in these elections. On January 1, 1907, the most significant referendum in Ontario history was to be held in the 19 municipalities that had made provisional contracts with the infant HEPC, in defiance of the controlling private power interests. An intense campaign was conducted by both sides in the last few months of 1906. Beck took personal command of the pro-public campaign, travelling tirelessly from one Ontario community to another, preaching the Hydro gospel. The private power interests, led by the Syndicate, put up a well-financed campaign. They continued their ridicule and vilification of Beck and "the dreamers," "visionaries" and "crackpots" who followed him. Beck, they said, had based his cost estimates on "cooked figures." The local utilities would be ruined if they were forced to sell power at the ridiculous prices promised by the power minister. Beck and Smith publicly refuted these charges at well-reported meetings.

Despite its best efforts, private power never had a chance. All 19 cities voted overwhelmingly to contract with the new "People's Power Company." In Toronto, the vote was 11,026 to 2,907. The other cities voted likewise: Berlin (Kitchener), Brantford, Galt, Guelph, Hamilton, Hespeler, Ingersoll, London, New Hamburg, Paris, Preston, St. Mary's, St. Thomas, Stratford, Toronto Junction, Waterloo, Weston, Woodstock. Yet the fight was still not over. In the referendum, most municipalities had not specified a maximum contract cost of HEPC power in their proposed bylaw. Toronto was one of the few cities that had. In most cities, people had simply voted for the principle of public power, based on their belief that the power would be more affordable and reliable.

Following the decisive New Year's votes, the Legislature authorized a 4 percent loan to the HEPC to begin construction of the new transmission system. But the bill required Hydro to provide every municipality with a firm maximum contract price and other terms that would be voted on at a second referendum on January 1, 1908. The bill also allowed municipalities to put to a vote the issuing of municipal bonds to construct a local distribution system, even where one already existed. This naturally alarmed the privately owned utilities. Their days were numbered should these second votes succeed.

Beck and his team worked feverishly to negotiate power supply sales with the private generators and to come up with accurate transmission system construction cost estimates so that the municipal contracts could be written well before the referendum date. Meanwhile, throughout 1907, the private interests redoubled their efforts to discredit the HEPC and, through sheer (and shrill) repetition, were having some effect. As late as October, many voters were uncommitted, moving back and forth in concert with the most persuasive argument they had last heard or read. It was, by all accounts, an epic battle for the hearts and minds of Ontarians.

Private power advocates pulled out all stops. They knew that the money spent on an anti-public campaign would be lost anyway if the HEPC won the local votes. Billboards warned of the perils of public ownership. Newspapers editorially opposed to public power were generously rewarded with paid advertisements ridiculing Beck. In Toronto, canvassers employed by the Syndicate went from door-to-door extolling the virtues of private power and disputing the claims of public power advocates. The Anti-Hydro Citizens Committee of Business Men warned that bankruptcy was certain if Hydro won the referendum. Incompetent politicians would be in charge of our electricity, they said, a recipe for disaster. The anti committee also spread the false notion that taxpayers, not ratepayers, would pay for the costs of any new municipal distribution system.

Sowing confusion about the distinction between taxpayers and ratepayers was, and still is, a popular tactic of anti-public power elements including the current Conservative government. For example, Conservative Energy Minister John Baird often says that the Hydro debt is such that a child born in Ontario today inherits a Hydro debt of $3,000. I have to presume that this is a deliberate distortion. The only alternative is a minister of energy who does not understand that the debt is paid down through electricity rates and that most electricity in the province is purchased by business and industry. The only way that Baird's imaginary child would be liable, as a future taxpayer, for that $3,000 would be if no more electricity was being sold in Ontario. Admittedly, the child will have to pay his or her share of the debt over time through Hydro rates, but if Baird thinks that private power companies have no debt that they too must recover through rates, he has been poorly briefed by his officials. All regulated power prices, public or private, are largely determined by the cost of debt financing, which is the reason public power is less expensive.

American private power interests had been involved in the anti-public campaign even before the 1907 referendum. They viewed the unfolding of the municipal movement with growing concern, afraid that this "public power infection" would spread south. Better, they thought, to nip this movement in the bud in Canada instead of waiting for the inevitable fights on their home turf. It is not known how much American money was spent in Ontario to try to defeat public power, but it is clear that U.S. private power interests were behind the February 28, 1907 resignation of Cecil B. Smith, the Commission's Chief Engineer.

The money of the anti-Hydro forces, however, proved no match for the passionate supporters of public power in every city, large and small. They too went from door-to-door: students, trade unionists, small business owners, engineers, even women who would not have the right to vote until a decade later in 1918, all hammering away at the fundamental idea that power should be provided at cost, that the wealth of nature belonged to us all. In the 1908 New Year's Day elections, Toronto, Berlin, Galt, Guelph, Hespeler, London, New Hamburg, Preston, St. Thomas, St. Mary's, Stratford, Waterloo and Woodstock all voted overwhelmingly for the municipal utility bond issue and to buy all their power exclusively from the HEPC, as soon as it could supply it. Ingersoll marginally voted against its own municipal utility, a decision that was later reversed in a subsequent vote. Brantford, Paris and Hamilton had relatively low-cost contracts with the privately owned Cataract Power Company, which had generated power since 1898 from DeCew Falls near the Welland Canal. But as it became known that Cataract's price was more than five times the cost of generating its

Syndicate partner Henry Mill Pellatt was the vivid embodiment of private power greed. In 1914, he built Toronto's Casa Loma, then the largest private residence in North America, for what today would be $60 million. He lost his fortune when public power triumphed. He died, penniless, in 1939. *City of Toronto Archives*

power, these cities moved over to the HEPC once their contracts expired. Cataract later sold out to Hydro and the DeCew Falls generating station still produces power to this day. It is, like the more than five dozen other hydroelectric stations in the province, an extremely valuable public asset that an NDP government would not ever consider selling.

While the average quoted price from the HEPC was $22 per horsepower per year, the contract power prices voted on by each city varied according to factors such as population and distance from Niagara. It obviously cost less per customer to service Toronto, at around $18/hp (slightly more than Beck had estimated two years earlier), than it did tiny New Hamburg, at about $29/hp. And while the big cities did not mind some level of subsidization to bring affordable power to smaller communities, they were unwilling to deprive their own citizens and industry of the greatest part of the benefit of public power. This compromise established a modified "user-pay" structure for Ontario public power that persists to this day, particularly in the local distribution sector.

Beck emerged from the 1908 referendum completely triumphant

and lost no time in ensuring that the promise of public power was fulfilled as rapidly as possible. The Syndicate, however, refused to be governed by something so intangible as democracy and continued to fight Hydro's plans for a transmission line from Niagara to Toronto. Such a line would be in direct competition with its own, which was already in operation.

The Syndicate went right to Premier Whitney to warn him that he was being duped by the ridiculously low estimates for the Toronto transmission line. Beck's engineers had estimated it would cost $3.5 million. In the meeting with the Premier, the chairman of the Canadian Bank of Commerce, Sir Edmund Walker, backed up the Syndicate's estimate that it would cost over $12 million, if it could be done at all. The HEPC plans called for a 110,000 volt transmission line at a time when 60,000 volts, which was what the Syndicate's lines carried, was widely regarded as the upper limit of transmission capability. The technical dispute was of great economic significance. If the Hydro wires could carry nearly twice as much power as the Syndicate's, the latter would quickly become obsolete.

Whitney worriedly posed these issues to Beck, who backed up his youthful engineers by saying he would take personal political responsibility if they failed to deliver. His faith was not misplaced. The project ultimately cost less than the original estimate of $3.5 million, despite the continuing efforts of the Syndicate to sabotage it. Among other tactics, Syndicate operatives privately approached the farmers along Hydro's proposed transmission corridor and warned them of the alleged dangers of the untested 110,000-volt lines, urging them to demand more for their right-of-way. The tactic worked. Hydro later calculated that it had been obliged to pay about twice what then-current market prices would have dictated for the rights-of-way.

When finished, the 110,000-volt lines worked well and quickly became the global standard for transmission network construction. At the same time, the Hydro engineers designed a new type of insulator, developed steel-core aluminum cable and improved tower structures, all of which eventually became industry standards. These innovations, and many more to come, firmly established Ontario Hydro as a world leader in power system technological development, in stark contradiction with the notion that public enterprises are necessarily bureaucratic, uncreative and non-entrepreneurial.

On November 18, 1908, work on construction of HEPC's transmission system officially began. It is hard to imagine that the same work today could have been done much faster, even though the construction crews at the time relied mainly on actual horsepower to haul the heavy materials through the fields and dig the holes for the tower foundations. Less than two years later, hundreds of miles of HEPC transmission lines began delivering public power to Ontario communities despite the stepped-up efforts of the now-desperate private power companies to stop what they continued to call this "public ownership madness."

A flurry of lawsuits, almost all sponsored in one way or another by private power interests, was unleashed on the HEPC, as well as on contracting municipal governments and the Province itself. The plaintiffs variously claimed that their "right" to choose their power supplier, even at a greater price, was being violated, or that the government's actions amounted to de facto expropriation, or that the municipalities had no right to get power from the Hydro if the final contract diverged in the slightest with what had been approved by referendum of local ratepayers. In a landmark decision in one such case, Sir John Alexander Boyd, Chancellor of the Ontario High Court of Justice, declared:

> The transmission and storing and distribution of electrical energy necessitates a system of control and regulation for the interests of public and private safety... The self-interest of the few must give way to the common interests of the whole body of incorporated inhabitants represented by the votes of the majority... The supply of light by means of gas or electricity, with the incidental advantages of heat and motive power connected therewith, appears to be a proper municipal function.

Encouraged by this decision, but unwilling to spend more years in court, the Whitney government passed the Power Commission Amendment Act in March 1909. The Act cleaned up the loose threads of the previous legislation at which the private power advocates had been tugging. The political fight then moved to Ottawa. Private power lobbyists, as active then as they are now, besieged the Laurier government with petitions demanding that it disallow the new Act on constitutional grounds. For months, the issue consumed Ottawa as corporate money flowed like the Niagara waters to the federal Liberals. At the same time, the Syndicate launched a massive and well-coordinated international media campaign aimed at damaging the credit of the Province and dissuading investors from putting their money into Ontario. Business journals in the U.S. and Britain, the chief source of Canadian capital at the time, were choked with articles and letters decrying the so-called "government-sponsored robbery." In mid-1909, the Syndicate compiled the most eminently authored and vitriolic of these attacks in a booklet: *The Credit of Canada: How it is affected by the Ontario Power Legislation; Views of British Journals and of English and Canadian writers and correspondents.* Nothing less than the destruction of the Dominion was predicted unless Ottawa disallowed the Power Commission Amendment Act, which, it was ponderously claimed, "repealed the Magna Carta." Much of what was printed in the booklet was lies and distortions. Most of the rest was wild conjecture or mere repetition of the "oft-proven fact" that public enterprises were inherently less efficient than private.

Looking back on these last-ditch efforts of private interests to kill public power, I am struck not only by their open contempt for democracy, if it interfered with their profits, but also by their willingness to jeopardize the future economic development of Ontario and, indeed, of all Canada. By painting the Whitney government as "socialist," a demonizing term then just coming into vogue, the Syndicate and its allies hoped to economically punish the province by driving away capital. If the Dominion Government failed to stop this madness, who in their right mind would invest anywhere in this irresponsible and unreliable country?

Once again, despite their money and influence, the forces of private power were defeated, in both their aims. In April 1910, the Federal Minister of Justice declared that Ottawa would not disallow the Ontario Act since it related "to matters declared by the British North America Act to be within the exclusive authority of the Provincial Legislature." And having been amply warned of Canadian socialism, British capital continued to flow into Canada at an accelerated rate. By 1914 it had reached £500,000,000 per year, nearly four times what it had been when the Whitney government was first elected.

Of all the dates that mark the progress of Ontario's public power movement, none is more climactic than October 11, 1910. The downtown portion of King Street in Berlin (now Kitchener) had been repaved, including the block that ran past the historic Walper House, where the municipal public power movement had been officially launched eight years earlier. New street lighting poles and lamps had been installed; between them were hung strings of bulbs that had never been lit. Arched across King Street was an incandescent banner, also unlit. The Mayor and his retinue greeted at the railway station the train from Toronto that brought Whitney, Beck and most of the Ontario Cabinet to the most auspicious event ever to take place in this industrious south-

A triumphal parade through downtown Berlin welcomed dignitaries to the October 11, 1910 ceremony inaugurating the first flow of Niagara power over Ontario's publicly owned transmission system. Adam Beck (centre) is accompanied by his wife, Lillian Ottaway, to whom he was deeply devoted. Provincial Treasurer A.J. Matheson tips his hat to the cheering crowd as the parade passes under the soon-to-be lit electric banner arched over King Street. *Hydro One Archives*

ern Ontario city. Also present was the local Member of Parliament, William Lyon Mackenzie King, who would later become Canada's longest-serving Prime Minister. Pictures of the event show Whitney, Beck and other politicians being driven by car to the town's hockey rink, its largest indoor venue. (Automobiles were still rare in Ontario and were certainly out of reach to all but businesses or the very well-off, a fact that helped fuel Beck's dream of an electrified intercity public transit system—the "radials"—that had been taking shape in his mind for years.)

As dusk fell, excitement mounted in the darkened, crowded auditorium and among the throng outside. Suddenly, a hush came over the crowd. All eyes were on 18-year-old Hilda Rumpel, dressed in a long white dress and wearing a red sash that read, simply, *Ontario*. On her head was a crowning wreath of tiny electric bulbs. In her hands was a cushion that held a switch. Stopping in front of Whitney, she shyly curtseyed and presented the cushion. Everyone in the arena held their breath, mesmerized by the moment. All knew it was a turning point in Ontario history. Pivoting to his right, the Premier reached over to grasp Adam Beck's hand and, without letting go, used it to press the switch that lit Hilda's crown and flooded the arena and the surrounding streets with light that began as energy from Niagara, one hundred miles away. The arched banner across King Street also lit up, spelling out these simple, moving and utterly true words: "For the People." The cheers and celebrations went on well into the night. In the speeches that followed, Beck promised to continue the fight for public power "until the poorest workingman will have electricity in his home." Whitney recounted the relentless attacks by the private power interests "to

Calling it "the Most Historic Event Ever Held in Berlin, and Probably in Ontario" the *News Record* described in lavish detail every aspect of the switching-on ceremony and the celebrations that followed. *Hydro One Archives*

— 52 —

destroy our power legislation and render it impossible for this wonderful new force to be used and enjoyed by the people except on the terms laid down by private corporations and individuals." Then he turned to Beck once again and said: "We, his colleagues, can never forget his steady confidence in the result and the bravery and pluck with which he stood up against all attacks."

The celebration in Berlin was quickly followed by similar ones in other cities that had voted to join the Hydro family. Before the end of the year, Guelph, Hamilton, London, Preston, Stratford, Waterloo and Woodstock lit up with inexpensive public power. By the early spring in 1911, switches had been pressed in Dundas, Galt, Hespeler, Ingersoll, New Hamburg, Toronto, St. Mary's and St. Thomas.

The world's first publicly owned long-distance transmission utility was off to a roaring start. It is still the only transmission network anywhere that was created by a direct vote of the people it was meant to serve. The present Conservative government's efforts to sell even part of Hydro One to the private sector without a vote of the people was a shameful repudiation of the most democratic event in Ontario history.

4

Rural Electrification, Farm Modernization

It took decades to fulfill Beck's vision of bringing affordable power to all of Ontario's vast rural regions. The campaign began in 1912 with the "Hydro Circus."

It didn't take long after the first flow of public power to confirm that affordable electricity as an economic development strategy was working. Six months after Toronto switched on "the Hydro," electricity demand in the city had grown 400 percent. By 1913, more than three dozen municipalities had voted to join the Hydro family, including Windsor, more than 150 kilometres past the end of the existing HEPC transmission line at St. Thomas. The network's extension to this southern tip of the province was completed in August 1914, just before the outbreak of World War I. With its proximity to Detroit and its abundant supply of power from distant Niagara, it was inevitable that Windsor would become Canada's main auto manufacturing centre.

Elsewhere in Ontario, the story was similar as steam-driven factories converted to electricity at an accelerating pace. In 1912, the Steel Company of Canada in Hamilton built the world's first all-electric steel mill; its confidence that the government was committed to ensuring a reliable supply of affordable power was the deciding factor. In the northwest, at Port Arthur (now Thunder Bay), the HEPC began supplying power to the grain elevator industry, a vital link between the economies of Eastern and Western Canada. This power did not come from Niagara; an unbroken province-wide transmission network was still decades away. But just as it did in the south, Hydro contracted with private power owners and developers in the north to supply electricity that would be carried over publicly owned wires. In the east, in 1912, Brockville became the first sizeable community in Hydro's St. Lawrence system. This disrupted the regional private monopoly of the

Mackenzie Syndicate's Electric Power Company, which was selling electricity at nearly $50 per horsepower, more than double Hydro's provincial average price. By 1913, Hydro's Severn System was serving several communities in mid-central Ontario, from Barrie on Lake Simcoe to Collingwood on Georgian Bay. Collingwood used its new source of affordable power to modernize and expand its growing shipbuilding industry.

There is no doubt that Adam Beck was the motive force behind this swift expansion of the public transmission system. His critics called him "reckless" and "mad" for using so much capital, so quickly, to bring power to smaller and more remote communities. But there was a method to this alleged madness, which benefits us to this day. A central aspect of Beck's vision was to use public power for industrial decentralization. Beck himself had been a successful industrialist in London, a thriving city that competed for new investment with Toronto and Hamilton. He had also absorbed the industrial expansion aspirations of the many municipalities that had come together in the public power movement. Then, as now, there were strong feelings throughout Ontario that Toronto dominated the economic life of the province and Beck, not having come from "Hogtown," perhaps shared them. "What we want," he declared in 1910, "is that every small village be an industrial centre; small towns and villages with plenty of manufacturing, rather than great factory centres with slums and congested populations."

Investment poured into the province. Much of the increased flow of British investment in the years after Hydro's establishment in 1907 came to Ontario because of affordable power. Low prices and reliable power also attracted American investment, particularly in the manufacturing and electrochemical industries. Meetings of local boards of trade and trade unions alike resounded with expressions of optimism for the future. The pace of job creation and productivity gains matched the expansion of the HEPC. Unemployment was declining, wages were rising. The unprecedented economic boom was indisputably helped along, if not caused by, the coming of public power.

It seems obvious and inevitable to us now that demand for electricity would naturally expand very quickly. But in the early years of Hydro, the projections of demand growth proved to be well below actual experience. That there might be a brisk uptake of public power was not immediately taken for granted, least of all by Beck. Without empirical evidence, he could not afford to subscribe to the "Build it and they will come" theory.

For Ontario's young public power system, affordable power was a chicken and egg problem. Transmission networks and generating stations are very expensive to build and they are completely useless until they are completely built. Once they are built, however, the more power that is generated and delivered by those facilities, the cheaper each unit of power can be. The same principle applies to other infrastructure systems, such as municipal public transit. The more people use their city's transit system, the lower the cost per rider. The converse is equally true. If you build an electricity system capable of delivering three times as much power as is actually taken, each kilowatt hour will necessarily cost more than it would if the system were used at capacity. A bus or streetcar that is only one-third full costs more per rider than one that is completely full.

Beck and his planners faced this classic business problem: Would electricity demand rise to consume all power that could be produced and delivered? If yes, then build away—they will come. The power will all be purchased, and at an increasingly lower cost for the consumer. But if there is a limit to demand growth, as there must certainly be, what is that limit and how far ahead of demand should the infrastructure to meet it be built? This is a critical economic question that we still face today. Excess capacity in a power system must be paid for; there's no magic here. But, on the other side of the equation, the cost of power shortages can be very heavy indeed, especially in an unregulated marketplace, as consumers in California, Alberta and, most recently, Ontario, have found out.

This is yet another reason why regulated public power systems continue to be, as they have been since Beck's time, better for society than unregulated private power systems. Now, as then, profit-seeking power producers quite naturally want to keep the supply/demand balance in their favour. They have no interest in lowering the cost of power. Why should they? In the absence of regulation, they will naturally charge as much as the market will bear. They can't be expected to give away electricity for less than consumers are willing to pay; that wouldn't be rational economic behaviour. Nor will they slit their own economic throats by investing in more production capacity than the market needs. A large corporation can afford a certain amount of underutilized capacity that it can ramp up to take advantage of rising demand. But small generators can't last long with unproductive capital assets. And a deregulated private power industry as a whole, even if comprised entirely of large companies, couldn't afford to have too large a surplus of generating capacity. As profits fall, capital takes flight. It's no use telling your shareholders that their sacrifices are good for the

In 1912, Waterloo public utility workers pose atop distribution transformers they had just installed. (L-R) Burt Warner, Jim Walker, Bill Reiber, Eby Rush were known locally as the "Squirrel Gang." *Hydro One Archives*

rest of the economy. They take no comfort in the fact that lower profits benefit their customers, most of whom are non-shareholders. In an unregulated system there is an inherent conflict of interest between customers and shareholders.

If all the customers of a utility are also its shareholders, however, there is no conflict of interest between the two. This is the case in a publicly owned system: lower power prices are functionally the same thing as dividends. Any "profits" stay in the system and are used to keep rates affordable, retire debt, maintain reliability and promote economic development. Public ownership and regulation affords a utility much longer capacity planning horizons than can be tolerated by competitively driven, privately owned power companies. Bottom line: in a privately owned electricity system power shortages are usually good for shareholders because supply shortages mean higher prices thus more profit; in a publicly owned system, they're bad for shareholders, who lose because of the shutdowns and production losses that come with the electricity brownouts and blackouts, on top of the higher prices. This is why public power systems typically err on the side of having a reasonable power surplus when planning capacity expansion, an issue that arises often throughout Hydro's history.

Despite the inherent economic advantages of public power, the early HEPC was in intense competition with the incumbent private power companies that already had customers and capital assets in place. This competition was ultimately resolved, in most communities, at the ballot box by ratepayers voting for specified maximum public power prices. For the HEPC to meet its price commitments, it had to keep construction costs low, negotiate good prices on power supply contracts and hope that demand would grow so that the unit cost of its power could continue to fall over time. No armchair visionary, Beck worked night and day to help ensure that demand for non-profit power grew quickly and could be met.

As a missionary, Beck was indefatigable, doing the work of three people. When he wasn't overseeing his growing cadre of young, dedicated engineers or negotiating power purchase agreements with private generators (or fending off their relentless attacks on the HEPC) or managing the prodigious number of political relationships needed to sustain the momentum of the public power movement for the four years (1907-10) before the HEPC could actually deliver any power, Beck took his vision directly to every potential customer who would listen. No village seemed too small for him to visit and preach the benefits of the people's power company. Indeed, by the summer of 1912, no farm was too small to be passed by in Beck's search for customers.

The Hydro Circus—likely named by Beck himself, given his customary marketing flair—was reputedly the world's first government-sponsored campaign to systematically bring electricity to large rural areas. It made a huge impression, not just in Ontario but also in the United States, where it helped inspire the U.S. rural electrification campaigns that came along some 20 years later as part of President Franklin D. Roosevelt's New Deal. More important for Ontario, it fired the imagination of rural voters that they could share in the electricity-driven prosperity of the cities. Beck knew that it would be a very long time before it would be economically feasible to build a complete electric utility infrastructure in rural Ontario. If it was to happen in the foreseeable future, it would have to be publicly subsidized. The Hydro Circus was a brilliant and successful strategy to make rural electrification a political issue.

Beck outfitted and dispatched two marketing teams. Each was made up of two covered wagons, pulled by horses that had to be changed frequently, so heavy was their load. In one wagon was a motor and the cables needed to connect it to the alien-looking power lines and towers that were sprouting up overnight on the rural rights-of-way the HEPC transmission system had to cross between Hydro cities. In the other wagon were transformers that stepped down the high voltage power carried by the transmission system to the lower voltages of a distribution system so that it could be safely used in houses and barns. The covered wagons were later joined by a truck carting the equipment and

Adam Beck's "Hydro Circus" was the first known rural electrification campaign in North America, possibly the world. This three-ton truck carried a washing machine, circular saw and other state-of-the-art appliances to farming community demonstrations often conducted by Beck himself. *Hydro One Archives*

appliances that were demonstrated with enthusiasm by HEPC marketers, frequently Beck himself. Audiences were enthralled with the electric lights and flabbergasted by the washing, sawing, milking, cream separating and pumping machines that did the work of many people for pennies a day. (In the early years of electricity, inventing and marketing electrical devices was very much like the software industry of our day.) Cooking on the electric stove was reportedly the most popular demonstration, especially at the fall fairs visited by the Circus.

It is difficult for us today to imagine the impact of electricity on farming. Suddenly, literally out of the blue, farmers were freed from the complete dependence on human and animal muscle power that had defined the limits of agricultural productivity since its very beginnings. Men, women and children alike were instantly freed from whole classes of drudgery that had always been the inescapable lot of farm families. It was a social and economic revolution almost without parallel. Beck himself best expressed his legacy in this regard in a 1913 address to public power advocates:

> If I have helped to lessen the household cares of the housewife by making electricity her servant ... if I have helped the farmer to make life on the farm more attractive, to help keep the boys and girls on the farm, then I have not laboured, nor have you cooperated with me, in vain.

What Beck did not claim (uncharacteristically, for him) is that Ontario cities benefited at least as much as our vast farming regions from the rural electrification movement he pioneered. The resultant surge in farm productivity due to electricity led to increased yields and lower prices, even as farm incomes rose. Better nutrition then followed because people could now afford a wider range of foods, and more of them. Affordable electricity on the farm meant more and cheaper food for everyone. Undoubtedly Beck was right that electricity helped keep some children on the farm because they now saw a better future there. But it also allowed many others to leave for opportunities only available in the cities, as their labour was no longer vital to the survival of their family's farm. Electricity gave many young rural Ontarians—the parents and grandparents of so many of us—a chance to pursue their dreams for a better life off the farm. Those who stayed on the farm did so increasingly out of choice, which is how it should be.

It is doubtful that Ontario's rural electrification drive would have ever happened in an unregulated, private power system. Even in the U.S., where 70 percent of utilities are privately owned, rural electrification only happened through government intervention. Rural rate subsidies, an Ontario-pioneered concept in North America (which Beck might have picked up on a 1912 trip to Europe), have typically been a key aspect of electricity regulatory frameworks everywhere in the world.

Despite Beck's determination, however, rural electrification progressed very slowly. One reason was logistical. Hydro was building transmission lines to the cities at a blistering pace, but not fast enough to satisfy the demands of the municipalities that had voted to join the family. Hydro engineers raced around the province conducting feasibility studies for municipal utility bond issues and Hydro supply contracts. In the early days of the electricity industry, there was a shortage of skilled labourers and professionals. The cities were given first call on what was available, a fact frequently noted in rural newspapers. But the biggest impediment to rural electrification was its cost. There were so few customers in rural areas that the per capita cost of building the networks to serve them was many times that of urban customers. This

The Hydro tent, shown here at the 1913 Renfrew Fair, was a very popular exhibit at rural Ontario fall fairs. The electric stove was reportedly the household appliance most coveted by farm families. *Hydro One Archives*

inescapable economic fact created a philosophical divide between town and country in Ontario that helped sweep the United Farmers of Ontario into government in 1919. The UFO was the first governing party in Ontario that was not Liberal or Conservative. It would be 71 years before that electoral milestone was repeated with the election of a majority NDP government in 1990.

The town and country divide can be concisely summed up as: Power at Cost vs. Flat Rate Power. There is a huge difference between these two concepts. Power at Cost was the central theme of the municipal public power movement. It implies non-profit power, but also implicit in the phrase is the concept that there is indeed a cost to that power. How is that cost determined? For municipalities, the answer was straightforward. Windsor City Council, for example, could calculate the cost to build and maintain the local distribution network. Added to that would be Windsor's share of the cost of the transmission line that served many municipalities as it branched its way from Niagara Falls to the Detroit River. Those were the network costs. Add in the cost of the energy delivered through that network, as contracted with Ontario Hydro, and voilà, we have a firm figure for Windsor's "power at cost," which was approved by the city's ratepayers in a 1913 referendum. Further up the transmission line, Chatham, St. Thomas, London, Woodstock, Waterloo and Brantford could go through the same exercise and arrive at a different cost of power, based mainly on their distances from Niagara and the amount of power taken by each municipality.

But what if Windsor decided to expand its distribution network into the surrounding rural regions of Belle River, LaSalle and Essex? That's a lot of territory to cover—a lot of wires and poles and transformers, not to mention a lot of workers to install them. The network cost per customer would have to rise. And if the cost of the network was to be shared equally by everyone connected to it—through a flat rate charge applicable to all customers—those who lived in the more densely populated areas would be subsidizing those who lived on the surrounding farms. The "cost of power" to the city dwellers would go up in order to bring equally affordable power to the region's rural areas. Since the residents of these rural areas did not vote in or pay taxes to the cities, there was no political or economic reason for city residents to subsidize their country cousins' power costs in this way. For these reasons, local distribution systems typically went up to the city limits but seldom further, except by special arrangement with customers on the other side of the line.

Hydro's early attempts to aggregate enough individual farms as customers, in order to make serving them reasonably economic, were unsuccessful. Efforts then focussed on first bringing the small rural towns and villages into the Hydro fold through the same bylaw referendum process conducted in the larger cities. Electrification would, it was hoped, then spread outward from these mini-hubs over time, as increasing demand lowered the per capita cost of service. It still didn't work. Rates in the Grey-Bruce region in 1915, for example, had to be more than four times those of the Toronto-to-Niagara municipalities if both regions were to have their own power at cost.

To rural Ontario, the solution was obvious. A movement for uniform Hydro rates throughout the province sprang up during and just after World War I, echoing the municipal power movement of a decade earlier. Uniform rate proponents did their own economic analyses showing that a reasonable uniform rate of about $20 per horsepower was possible if only a handful of the larger industrial cities gave up their much lower rates of between $11 and $15 per horsepower. But by 1919, there were over 130 Hydro municipalities. A provincial uniform rate would require a revision of each one's ratepayer-approved contract, an unlikely prospect.

Beck very much approved of subsidies for rural electrification, but he shared the view of the young Municipal Electrical Association (an organization that survives to this day as the Electricity Distributors

Association) that uniform rates would destroy public power. Here was his reasoning: If St. Catharines' rates, for example, jumped by 50 percent in the cause of uniform rates, private power producers would then be able to significantly undercut Hydro's prices in that city. If, as a result, St. Catharines then left the Hydro fold, the HEPC's per capita system costs would rise, further eroding Hydro's competitiveness. Private power would simply cherry-pick the most lucrative urban markets and leave the unprofitable rural markets to Hydro. The only solution would be to outlaw private power altogether, which was deemed to be politically unacceptable if it resulted in higher rates in the cities. Ending private power was also economically impossible at that time, as it would have required buying out all major private power companies in the province.

The uniform rate issue loomed large in post-war politics in rural Ontario, but was largely ignored in the big cities and by the Conservative government. Throughout 1919, uniform rate associations in different parts of the province lobbied large city utilities relentlessly, but gained no support for their "revolutionary" idea. Hope came in the form of a stunning political upset on October 20, 1919, when the United Farmers of Ontario, who entered their first election without a designated leader, won 44 seats in the 111-seat Legislature. The Conservatives, led by Whitney's successor, Sir William Hearst, dropped from majority government to only 25 seats, two less than the Liberals. By allying with the dozen newly elected Labour members, the UFO was able to form a majority government in the same year as the Winnipeg General Strike. Great change was in the air.

Beck, who had been knighted in 1914, had been approached in early 1919 by a committee of concerned Conservatives who thought he would be a more popular leader than Hearst. Beck's price was a free hand with Hydro and the "radials," his scheme for an electrified intercity public transit system in southern Ontario. The committee demurred and Beck decided to run as an Independent. To everyone's surprise, Beck lost his long-time seat by a small margin. The UFO then privately approached Beck and offered to make him premier. His conditions were the same as he laid out to the Conservatives—a free hand in all Hydro matters. In the end, partly because they feared Beck's radials would consume government attention and money at the expense of farmers' needs (they were undoubtedly right) the UFO selected as their leader and new premier the 41-year-old E.C. Drury, a former Liberal turned agricultural activist. Drury offered Beck the post of Minister of Power, but Beck turned him down, as he had Hearst. He did not want to operate under the constraints of Cabinet government, an attitude that had been losing him friends and supporters at Queen's Park for many years. With his still-solid municipal power base, however, Beck felt he had no need to curry favour in the provincial government.

In May, 1920, Drury appointed a legislative committee to study the idea of the uniform rate and make recommendations. The committee was chaired by a UFO member of the Legislature, John G. Lethbridge, who was known as a supporter of uniform rates. Five months later, the committee concluded that the uniform rate was, in fact, impractical. Instead, it suggested a $2 per horsepower tax on all Hydro delivered to municipalities for less than $30 per horsepower. The millions that would be raised from this levy every year would subsidize the construction of rural transmission and distribution networks.

Reaction to the Lethbridge report was predictable. Urban utilities and industrial users opposed the tax, claiming it would make them uncompetitive. Rural Ontario embraced the idea as a just and almost painless way to redress the growing gulf in electrification between the cities and the farmers who kept them fed. Both sides lobbied the government furiously. Beck proposed a solution to this discord directly to Drury: Let the government commit $250,000 a year to subsidize one-third the cost of building Hydro lines in rural regions not yet served, a sum much less than the tax proposed by Lethbridge. Farm advocates were incensed and the lobbying raged on throughout the winter. Finally, in April 1921, the besieged Drury introduced legislation calling for a Hydro-Electric Extension Fund that would finance up to 50 percent of the cost of primary transmission lines into rural areas. Less than six months later, work began on the first subsidized rural lines into Wentworth County and a dozen or so other townships across the southern part of the province, up to Ottawa.

In June 1923, the Conservatives were decisively returned to power under the leadership of G. Howard Ferguson. One of the reasons they captured more than half the UFO's seats in rural Ontario was their promise to improve the "bonuses" for rural lines. In plain terms, the party promised to be more socialist than the UFO. Beck had allowed himself to be wooed back into the Conservative fold and won a hands-down victory in London. He then orchestrated the rapid passage of the additional bonus legislation, which subsidized half the cost of secondary transmission lines into farming communities. Taken together, these public subsidies put the cost of electrical service within reach of most rural residents and businesses, with minimal effect on city rates when their Hydro contracts were later renewed. In one form or another, strategies aimed at levelling the cost of power between Ontario's cities and rural areas continued for the rest of the century.

In the U.S., meaningful government assistance to get power to the farms only began in 1935 with the Roosevelt-era Rural Electrification Administration (REA) and the Norris-Rayburn Act, which provided 25-year loans for rural line construction. The fact that rural America received loans, not grants, for line construction meant that rural Ontario rates stayed much cheaper. But the early New Deal spirit injected more energy into rural electrification than had been seen in Ontario since the Hydro Circus and, in fact, several travelling exhibitions promoting the benefits of electricity to farmers in almost every state were clearly the descendants of Beck's brainchild. The REA helped farmers set up their own electric cooperatives (hundreds still exist) and even assisted in the financing of farm appliances and equipment, which built

The travelling Hydro rural exhibit at the 1946 International Plowing Match held at Port Albert, on Lake Huron, near Goderich. Ontario's complete rural electrification was still more than a decade away. Today, Port Albert is home to a co-operatively funded wind turbine, with more on the way. Renewable energy promises a whole new era of rural electrification. *Hydro One Archives*

up load demand. As a result of these aggressive New Deal strategies, rural electrification in the U.S. quickly overtook Ontario. By mid-century, over 90 percent of all U.S. farms used electricity; only 60 percent did in Ontario. It took Ontario another ten years before it could claim complete rural electrification. The American New Deal strategies to bring power to the farms were better and more comprehensive than Ontario's public programs. In neither country, nor in any other part of the world, did the private sector play any appreciable role in spreading rural electrification.

Just because rural electrification is complete does not mean there is no longer any need for public power in rural Ontario. Quite the reverse is true. Public power in the 21st century can and should be at least as beneficial to rural communities as it was in the last century. This goes way beyond merely subsidizing the maintenance and refurbishment of rural lines to keep them reliable. There is a whole new world of small-scale renewable energy generation technologies that are ideally suited to rural areas. Small flotillas of strategically located wind turbines could provide significant amounts of power to scores of farms and the rural towns they surround. Hundreds of small scale run-of-the-river hydroelectric sites could be quickly developed, many in the most remote areas of the province. Rooftop solar heat collectors, not nearly as costly as solar electricity panels, could provide water and space heating for individual barns and other farm buildings. Biomass-fuelled generators can create electricity from farm wastes, including animal manure. More energy efficient farm lighting and machinery could reduce electricity costs while making farms more productive. A publicly owned utility could greatly assist in the financing of such technologies and thereby help make farmers more energy independent. This would, in turn, reduce the need for more transmission and distribution investment, an economic and environmental benefit for all Ontario. Private generators and utilities would obviously not be interested in actually reducing rural power demand, let alone keeping rates affordable. For these reasons, public power will continue to be as important as ever for rural and Northern Ontario.

5

The Queenston-Chippawa Project

Comparable in scope to the Panama Canal, Hydro's construction of the world's largest generating station fuelled Ontario's post-World War I industrial development.

The Hydro-Electric Power Commission of Ontario began its operational life in December 1910 by delivering privately generated power through publicly owned wires networks. This made economic sense, as the private power concessions already in place at Niagara had considerable surplus generation capacity. Since they were losing their retail markets anyway, as municipalities voted in droves to do business with Hydro, private producers had little choice but to sell their output to the HEPC. In 1910, for example, the American-owned Ontario Power Company agreed to supply Hydro up to 100,000 horsepower, a huge amount of electricity at that time. By the end of 1914, however, HEPC demand on its Niagara system had grown to three-quarters of that supply at a time when manufacturers were being asked to convert their factories to full-time production of war materials. By Armistice Day, November 11, 1918, electricity demand exceeded 200,000 hp, nearly triple what it had been only four years earlier. At this rate, Ontario would run out of power within four more years.

Even before the war, it seemed inevitable that Hydro would ultimately enter the generation business, and not just because the boundless ambition of Beck, the Chairman, made that a safe bet. It was (and still is) much cheaper for a government-backed utility to finance the construction of generating facilities. Stable, democratic governments in prosperous countries command the lowest interest rates because lenders are confident that they will be repaid. Bankers may publicly grouse about government debt at a Chamber of Commerce luncheon, but will line up outside the Minister of Finance's door to get their share of that debt later that afternoon. Even small differences in bank lending rates to a hydro utility can make a large difference in the electricity rates that have to be charged to repay the loan. To illustrate, let's take a $10 million construction loan to be paid off over 30 years. At a government-guaranteed 5 percent financing rate, the total interest cost of the loan will be $9.3 million. At a corporate-guaranteed 8 percent rate, interest costs are $16.4 million, which is 76 percent more than the public utility—and its customers—would be paying under the lower interest rate. On top of that, of course, the private utility has to generate profit for its shareholders and pay (typically) bloated private sector salaries and commissions, all of which means still higher prices for its customers.

In July, 1914, the month World War I began, Hydro first entered the generation business with the purchase of the five year-old Big Chute station on the Severn River from a private company, the Simcoe Light and Power Company. It served the towns of Collingwood, Barrie, Midland, Penetanguishene and their surrounding districts. Three months later, the first Hydro-built generator at Wasdell Falls, also on the Severn River, began producing power for the Beaverton-Canning district. At the station's opening, Adam Beck said: "We are off now. Just let anybody try to stop us."

In that same year, work began on Hydro's second self-built hydro-electric station at Eugenia Falls on the Beaver River, near Meaford. It began serving the Owen Sound area before the end of 1915. It is instructive that Eugenia Falls and Big Chute are still producing power, and pollution-free power at that, nearly 90 years later. By contrast, the design lifespan of a nuclear or coal station is 40 years, at best.

Four major steps towards fully public power were taken during the war:

In 1916, the government bought the Central Ontario system of the Mackenzie Syndicate's Electric Power Company: stations, wires, poles and transformers. The old Beck rival, unable or unwilling to come up with the capital needed to upgrade service in the Oshawa and Peterborough areas enough to support the war effort, finally threw in the towel and retreated from the east of Toronto region. (The Syndicate held on to its Niagara generating station until 1922, but Beck lived to see the day of his complete triumph over private power.)

In 1917, the HEPC bought out its chief Niagara supplier, the Ontario Power Company, and immediately began to upgrade its generating station to squeeze out 45,000 hp more of desperately needed power.

Also in 1917, Sir Adam Beck and Sir William Hearst triumphantly broke ground for the Queenston-Chippawa Generating Station at

Niagara, then the most ambitious hydroelectric power project in history.

In 1918, Ontario Hydro began construction of the first publicly owned hydroelectric generating station in the province's Northwest, at Cameron Falls on the Nipigon River.

Hydro development in Northern Ontario was just as dramatic as it was in the south. At the top of Lake Superior, the cities of Port Arthur and Fort William (now Thunder Bay) were booming, drawing people and money from the southern part of the province and many immigrants from Europe looking for a better life and willing to work hard for it. New grain elevator terminals crowded the waterfront; the pulp and paper industry couldn't keep up with demand and mineral exploration in the region promised great riches just beneath the soil. Although Hydro had been serving Port Arthur since 1910, it had been doing so with power purchased under contract with the privately owned Kaministiquia Power Company. That contract was due to expire on December 21, 1920. The 10,000 hp called for under the contract was almost completely absorbed by 1916. The cities faced an abrupt halt to their economic development and prosperity without more power. City councillors implored Hydro to get into the generation business and Beck agreed, provided the ratepayers were supportive. The January 1, 1917 referendum vote in both cities was nearly unanimous. Hydro contracts were signed and by year end the provincial government had formally approved the new projects.

Premier Hearst, who was from Sault Ste. Marie, probably had a bet-

William H. Hearst (1864-1941) was born in Arran Township, south of Owen Sound, but lived much of his life in Sault Ste. Marie. He became premier upon Whitney's death in 1914. He was a supporter of Beck's, and ordered the 1917 referendum that recorded overwhelmingly popular support for publicly owned generating stations in Ontario. *Hydro One Archives*

ter idea than Beck of how physically challenging these ambitious projects would be. The crews had to build roads through trackless, uninhabited forest and live in small, very basic dwellings throughout most of the year. Depending on the season, the workers (and later their families) could be found either being eaten alive by blackflies and mosquitoes or numbly shovelling snow off their equipment and partly built structures so that they could continue to work until 20-below temperatures finally drove them away for the rest of the winter.

In June, 1919, hand drilling and excavation of the often-cursed Canadian Shield began. A month later, workers were shovelling out, again by hand, the channel that would funnel the water from Lake Nipigon through the generators. At one point, nearly 700 men worked at the site. The engineering feats matched the physical endurance of the workers, a great story in itself. At precisely midnight on December 20, 1920, the Kaministiquia contract expired and totally public power started flowing from Cameron Falls to Thunder Bay, 80 miles away. In one of the most difficult terrains and climates on earth, the still-learning Hydro team, every one of them a public sector worker, had constructed in the wilderness a state-of-the-art power plant, on time and on budget.

It was the first of ten major generating sites in the northwest built by Hydro over the next four decades. Together, these sites can generate over 1,100 megawatts and have fuelled an enormous amount of economic development in my part of the province. That's the good news. The bad news is that these and other Northern Ontario power developments often trod on the ancestral homelands of First Nations peoples, with the tacit consent of the government.

In recent years, Ontario Hydro made efforts to redress legitimate First Nations grievances, which are many. First Nations leaders have sought to arrive at agreements without involving the courts, a commendable approach that has resulted in many compensation agreements. Progress has been made, but we will not have paid the true cost of public power in the north until all of these historical wrongs have been fully resolved.

The story of Hydro's Queenston-Chippawa power project at Niagara began years before the groundbreaking ceremony in mid-1917. In early 1914, Beck sent the first team of surveyors and engineers to the area to investigate the feasibility of building a massive power station downriver from the Falls. The idea was to take full advantage of the 100-metre drop in water level between Lake Erie and Lake Ontario. The existing private power plants had located their turbines partway down or very near the face of the Falls, harnessing only a fraction of the potential energy in the water that tumbled from the Niagara River above to the Gorge below. By diverting water at upstream Chippawa and funnelling it to a power station at Queenston, five kilometres past the Falls, the full power of the forces of nature could be captured and converted to electrical energy. As an important bonus, there would be no unsightly powerhouses cluttering the view of the magnificent Falls. It would be a project of epic proportions and therefore right up Sir Adam Beck's alley.

Beck did not have the authority to either construct or purchase any power generation assets. These all had to be approved by a Cabinet to which he had not belonged since the 1911 election, when he turned down Whitney's invitation to be Minister of Power. The purchase and building of small stations on the Severn River had been easily approved, but a mega-project of the scale envisioned by the less-than-diplomatic Chairman was a different matter altogether. Beck had to lobby relentlessly to get the dying Whitney's support before the 1914 election, and then had to do it all over again the following year to Whitney's successor, Premier Hearst. Beck said the project would take three years and cost $20 million. Hearst agreed with Beck that more power would be needed soon, even if the war in Europe did not drag on (which it did, for three more blood-drenched years). But such a financial commitment should, because of its size and permanence, be subject to a vote of the people who would ultimately be paying for it.

Wouldn't it be nice if today's Conservatives had the same democratic values as their party had back then?

For his part, Beck welcomed another municipal referendum, believing it would be a walk in the park compared to the intense ballot box battles with private power he had waged and won in 1907 and 1908. It was not to be that easy. Throughout 1916, the Mackenzie Syndicate, and other private power interests, repeated their earlier campaigns, recycling the claims that government enterprises are "naturally" less efficient and that public debt would surely soar to unsustainable levels. They admitted that publicly owned wires networks were, in fact, bringing down the cost of power in most places, but they argued this was a lot different than the government attempting to build the largest generating station in the world using untried engineering techniques and requiring unprecedented organization. Besides, hadn't private power owners shown in recent years that they could work with the HEPC to bring affordable power to Ontario? The risks of Beck's scheme were completely unnecessary, they said, a position taken up by many eminent financiers, though not by many energy-hungry industrialists.

Predictably, the intense ideological fight energized the now 60-year old Beck. Once again, he travelled to virtually every city in which the plebiscite would take place, engineers and accountants in train. Take the next logical step, Beck told his audiences. The HEPC has proven its engineering and organizational competence and financial responsibility. It is simply a fact that reserves of private power are declining. Demand will overtake supply sooner rather than later. We cannot afford to put Ontario's energy future completely in the hands of privately run corporations and allow them to determine how much power we have. We know for certain that they will not build new generation capacity unless it is assured of being highly profitable. The people of Ontario have already crossed over the bridge to public power. We must not turn back.

Once again, the money of the private power interests was no match for Beck's impassioned eloquence and organization, magnified, as before, by an army of public power advocates in every municipality. On January 1, 1917, ten years after the original public power plebiscite, a resolution authorizing municipalities "to develop or acquire, through Hydro, whatever works may be required for the supply of electrical energy or power" passed overwhelmingly in more than 100 cities. Hearst quickly introduced a bill authorizing the Queenston-Chippawa project. Because Beck had been preparing for a year as if approval was certain, work on the immense power plant, and the canal that would supply it, was able to begin less than four months after the legislation passed. More impressive than its size was the fact that Hydro engineers had designed a system that extracted from the falling water two to three times the energy that the same volume of water was producing at existing private power stations at the Falls. It was an historic breakthrough in supply-side energy efficiency that has saved the province billions of dollars in energy costs over the last 80 years.

The project's size and complexity made it comparable to the building of the Panama Canal, which had opened in mid-1914. The water of the Niagara River had to be diverted above the Falls at the mouth of the Welland River. The flow of the Welland had to be reversed. The water was to travel through a 6.5-kilometre channel to a deep, concrete-lined, 8-kilometre-long canal. There the water would be fed through a series of wide tubes, called penstocks, to ten generating units, each of which would produce up to 65,000 hp. The powerhouse would be as tall as an 18-storey building and as long as two football fields. Over 100 kilometres of railroad track had to be laid. Millions of cubic metres of soil and hard rock were excavated. Geological surprises frequently stalled the

A specially built electric train was used to carry away countless tonnes of earth that had to be removed during the construction of the Queenston-Chippawa project at Niagara in 1918. *Hydro One Archives*

work until new plans and calculations could be made. Engineering obstacles were overcome on an almost daily basis. Thousands of workers had to be housed, fed, equipped and organized. Materials in unheard-of quantities had to be produced and deployed. Machines of unprecedented size and function had to be designed, built, transported and installed. And all of this in the midst of a war thousands of miles away that was draining away for cannon fodder an already limited workforce and voraciously consuming the capital and expertise needed to keep the Queenston-Chippawa project on schedule.

Within months, it was clear that Beck's $20 million estimate was going to be woefully insufficient, even though it had been based on more than two years of careful studies by several teams of engineers, some of them retained by Hydro from the University of Toronto. Hydro periodically issued upwardly revised estimates which looked for all the world as if the anti-public power forces had been right. By the end of the war, Beck's dream was being labelled a nightmare by private power interests and their political and media allies. Beck took the heat and pressed on, visiting the site frequently and personally exhorting the workers to give their all for what would most certainly be a perpetual monument to the greatness of Canadian ingenuity and determination. Then he would rush back to Queen's Park to allay Hearst's fears that he had staked his political legacy on a white elephant.

With the return of the soldiers, the pace of work picked up. At the peak of construction, over 10,000 workers were employed, but at twice pre-war wages due to galloping post-war inflation. That inflation also pushed up the price of materials and machines. Add in the unexpected engineering challenges and the enlargement of the initial project from 100,000 hp to more than 500,000 hp (which ultimately proved to be a wise decision), and the final cost of the project came in at a staggering $73 million. This would have made it the Darlington nuclear fiasco of its day, except that the Queenston station was producing reliable, clean, low-cost power for 70 years before Darlington opened and will likely still be producing that power seventy years after Darlington is decommissioned at great cost and buried under an artificial, uninhabitable mountain.

In 1919, Hearst asked Beck to bring in respected outside engineering consultants to review the project and make any recommendations they felt necessary. Before that happened, however, the UFO won the October election and Beck suddenly had to answer to an unknown Premier.

E.C. Drury had only been chosen for the job of premier because Beck himself had been unable to come to terms with the UFO leadership. During 1920, two separate groups of independent consultants studied the project. Though there were criticisms of various aspects, many of them due to 20/20 hindsight, a joint team of American and Canadian hydroelectric engineers concluded in the fall that Hydro's design was sound and that the cost overruns were not its fault.

Drury remained dissatisfied with the cost of the project, but that didn't stop him from attending the opening of the first generating unit on December 29, 1921 and switching on the huge banner of lights that read: "The Largest Hydro-Electric Plant in the World." The station was then 18 months behind schedule. How much of this delay was due to the war can only be guessed at. What is now known, however, is that Beck's prediction of power shortages came true and that the project later named the Sir Adam Beck Generating Station, proved essential to Ontario's economy. As early as 1919, Hydro had to begin curtailing contracted power to municipalities during times of peak demand, a troubling trend that did not go unnoticed by potential investors. The mammoth new station, despite its great cost, came on-line just in time to inject new energy, literally and metaphorically, into the province's continuing industrial development.

Thirty years after power began flowing from the first Sir Adam Beck station, the Sir Adam Beck 2 came on-line. Built right alongside Sir Adam

The Queenston-Chippawa plant, later renamed the Sir Adam Beck 1. When commissioned in 1921, it was the largest hydroelectric station in the world. It still produces power today, even more than it did when first built. *Hydro One Archives*

Beck 1, the new station produced three times as much power, taking its water from two huge 8-kilometre-long tunnels cut through the rock below the city of Niagara Falls. In another of their finest moments, Hydro engineers also designed a 750-acre storage reservoir behind the powerhouse. During off-peak times, when most of us are asleep, some of the generating units are used to pump water into this reservoir, which is then used later in the day to generate even more energy during peak hours, at extremely low cost.Today, 50 years later, this "double-duty" concept pioneered by Hydro avoids significant greenhouse and acid gas emissions, since coal is typically used to supply peak power needs that cannot be met by hydroelectric power. It also ensures that the flow of water over the Falls is at its greatest during daylight hours, particularly during the peak tourist season. The original idea that indirectly led to public power—preservation of the natural splendour of Niagara Falls—was always honoured by Ontario Hydro.

Today, the combined Sir Adam Beck stations produce nearly 2,000 megawatts of eternally renewable, emission-free power, more than ten percent of Ontario's typical daily demand. The cost: about half a cent per kilowatt-hour! Why would any government in its right mind think for even a moment of selling off such an incredibly productive and valuable asset of the people?

6

The Radials Controversy

Three Royal Commissions in three years diminished Beck's influence and proved fatal to his dream of an intercity electric rail network.

Sir Adam Beck's last four years on earth were not his happiest. The great achievement of the world's largest power project at Niagara continued to be marred by persistent criticism of its cost overruns. Resistance to his vision of an intercity electrified rail system was growing. Calls to rein in the Chairman, who frequently acted as a power unto himself, echoed constantly in the Legislative Chamber and in the UFO Cabinet. Having brought affordable public power to Ontario, Beck was now accused of going too far, too fast, too independently. His unique public stature and reputation began to erode as the political memory of what it was like to live with unreliable and expensive power, or with no power at all, began to fade.

In early 1922, shortly after he had lit the sign over the Queenston generating station, Premier Drury appointed a well-known Toronto lawyer, William D. Gregory, a Liberal, to chair a five-member Royal Commission, the Hydro-Electric Inquiry Commission. Its principal mandate was to investigate the reasons for the gap between the original estimates of the Queenston-Chippawa project and its actual costs, which were still growing. The commissioners were told to go beyond this, however, and look into everything the HEPC did, including its administration. Beck was incensed at this intrusion and his many allies in the Municipal Electric Association were dismayed as well.

As much as I admire Beck, I would have done the same thing had I been in Drury's shoes. There was plenty of evidence at the time that Beck did not always see himself and Hydro as subordinate to the Legislature. I can understand that it must have seemed tediously frustrating for Beck to have to endlessly explain the complex economics of power systems to politicians who didn't understand the difference between watts and volts, but who were easily alarmed, simplistically, he thought, by numbers with lots of zeros. Beck sometimes forgot that democracy must be paramount. All public expenditures must be under strict public control and subject to accountability. There can be no exceptions to this rule, because if there is one, there are many. While there are inevitably some government expenditures that must, for unimpeachable reasons, be kept confidential, there must be political mechanisms in place to ensure that such confidentiality does not mean reduced public accountability. Beck's use of Hydro funds to promote the radials was wrong. So was his habit of not consulting his fellow commissioners on many decisions, even major ones. So were his lengthy delays in reporting some expenditures. Beck *had* to be reined in, just as his successors who created the Hydro nuclear cult 50 years later had to be made accountable and brought under control.

I can easily imagine today's opponents of public power saying: "See, this is what we're telling you about the inefficiency of public power. Give someone the temptation of the public purse and independent Commission status and you can pretty well guarantee that somewhere, sometime, something will go haywire." Fair enough, but so what? If that observation is meant to suggest that all power (lower case "p") corrupts, the observation is no doubt true, but certainly trivial in terms of telling us something that we didn't know already. If, however, the observation is meant to suggest that accountability is somehow easier for private companies, then the claim is indeed exciting, but, alas, false. There is not one whit of evidence to suggest that private control of public power resources, especially in a deregulated environment, will not at some time go off the rails in an even more spectacular fashion than was ever witnessed in Beck's day. The evidence to the contrary mounts every day.

The Gregory Commission was very thorough. Four interim reports had been submitted to the Legislature by May 1923. They contained some criticism of Hydro accounting methods and Beck's use of general Hydro funds to finance radials in Essex County and Guelph. There were to be more interim reports, including the one on Queenston-Chippawa, but a June 1923 election intervened. Hoping that the Commission would die along with the UFO government, Beck successfully ran again as a Conservative for his old seat in London, vowing to defend Hydro with his last breath on the floor of the House.

The UFO-Labour coalition was indeed crushed by George Howard Ferguson's revitalized Conservatives, but this did not produce Beck's desired result. Ferguson, who saw highways, not radials, as the key to Ontario's transportation future, and who wanted to rein in Beck as much as Drury had, made only a half-hearted effort to restrict the Gregory Commission to an investigation of Queenston-Chippawa. Since its original terms of reference were not formally changed, however, the Inquiry pressed on. Its final report, tabled by Ferguson in the 1924 session of the Legislature, was an extensive and impressive piece of work and was widely reported in the press. Its findings contained both good and bad news for the Chairman.

The Gregory Commission unequivocally endorsed the principle of public power and praised Ontario Hydro's engineering department, calling Hydro's facilities "exceptionally well operated." The HEPC's accounting system was adequate and its organization was financially sound. Sir Adam Beck was given lavish credit for his role in building up Hydro and protecting it against those who would destroy public power. On the Queenston-Chippawa project, the Commission was clear: It was "a magnificent piece of engineering" and the cost overruns, though unfortunate, were not the fault of Hydro's engineers or administration. This finding was based on an in-depth review of the project by independent engineering consultants. Ontario Hydro, the Commission concluded, had been doing its job and doing it well. The Gregory Report effectively buried any vestige of hope of a return to private power in Ontario.

Ernest Charles Drury, (1878-1968). Co-founder and first president of the United Farmers of Ontario, Drury became the province's eighth Premier after the leaderless UFO allied with Labour after the 1919 election to form the first Ontario government that was neither Liberal nor Conservative. Drury was the first Premier to assert government control over Hydro, no small feat while Adam Beck was still in charge. *Hydro One Archives*

But it also buried, for a while at least, Adam Beck's reputation as the saint of public power. The report scorched Ontario Hydro's (i.e. Beck's) obsession with radials and heavily criticized Hydro expenditures on them. It also documented cases of Beck spending HEPC money without proper authorization and frankly reprimanded the Chairman for his lack of respect for the Legislature and his colleagues. His lowballing of project cost estimates seemed to be habitual. Indeed, the report said, no public official other than Beck could have gotten away with such conduct. It recommended a "complete revision" of the Power Commission Act.

Gregory also drew attention to the significantly lower rates on the Niagara system compared to the Central Ontario, St. Lawrence and Northern systems which did not receive power from Niagara. These differences arose from Hydro's fairly strict and narrow definitions of "power at cost." Those areas of the province not yet connected to the transmission grid emanating from Niagara Falls got no benefit from this resource that supposedly belonged to all the people. Although public power was still cheaper than private power virtually everywhere in the province, that of cities and towns wired to Niagara was by far the cheapest. This difference, the Commission noted, was drawing industry away from other parts of Ontario to the southwest, contrary to Beck's original vision of making every town and village an industrial centre. Some of this differential, the report said, was due to accounting practices that artificially kept Niagara system rates lower than they should be. Changes were recommended.

When he tabled the Gregory Report, Premier Ferguson said it, "completely vindicated the Hydro project," which was true. Beck, however, was not comforted by this. He spoke passionately in the Legislature for several hours in his own defense, taking issue with the report's findings on his conduct, especially the accusations surrounding his unauthorized funding of the radials. At an age when most people are retired, Beck still dreamed of an extensive system of intercity railways akin to those he had seen on his 1912 European tour.

So did many others in Ontario. The radials movement, because it failed, is now an historical curio. At the time, however, the idea of the radials had broad appeal throughout southern Ontario. They made sense. Prior to World War I, roads were uniformly terrible, where they existed at all. Travel was difficult and the cost of getting goods to markets very high, which inhibited economic growth. Railroads, on the other hand, were fast and reliable. Moreover, electric railways, including urban streetcar networks, were already well established in Ontario. By 1912, there were more than 2,000 kilometres of track devoted to them, mostly in the Niagara-to-Toronto region that was served by the

Mackenzie Syndicate's private transmission system.

In late 1912, the Municipal Electric Association passed a resolution calling for "a system of electric railways, including street railways, to be owned by the municipalities" and financed along the same lines as the electric utilities. In May 1913, the Ontario Legislature approved Beck's Hydro-Electric Railway Act. It gave authority to the municipalities, subject to government approval, to build and operate hydroelectric radials at their own expense. Later that year, delegates to a Toronto conference of municipal leaders formed the Hydro-Electric Radial Union of Western Ontario and asked Ontario Hydro to build the lines wherever needed. Similar groups sprang up around the province and, with Beck's fervent orchestration, relentlessly lobbied the provincial and federal governments. On March 25, 1914, a delegation of two thousand radial devotees travelled (by rail) to Ottawa to ask Conservative Prime Minister Robert Borden for help in financing Hydro radials. They went away encouraged by the unctuous Borden, but empty-handed. No federal help was ever extended. A smaller delegation to Queen's Park the following week won better results, no doubt because of the impending provincial election. The new Hydro-Electric Railway Act, 1914, repealed the previous year's Act of the same name. The old Act had required municipalities to finance their own radials; the new one gave Ontario Hydro power to issue its own bonds to help finance a province-wide radial system. It looked as if Beck's scheme was well on the way to realization.

The coming of the Great War took some of the steam out of the radial movement, as resources and power were needed for the war effort. Beck himself was undeterred. He persuaded several municipalities to hold plebiscites to approve local radial projects and these votes almost always resulted in large majorities in favour. And in a repeat of the grand 1906 march on Queen's Park, 1,500 municipal representatives paraded up University Avenue in March 1915, to demand of Premier Hearst's government a radial subsidy of $3,500 per mile. Hearst was unwilling to consent to such an open-ended commitment and said he needed more time to study the issue. That delay ultimately proved fatal to the radials. A 1916 amendment to the Hydro-Electric Railway Act prohibited any radial bonds from being issued until the end of the war. By then, the railroad era was coming to an end. Use of the much cheaper and more flexible internal combustion engine was on the rise. When the U.S. entered the war in 1917, tens of thousands of trucks were mobilized to carry food and other essential supplies from the Midwest to the Atlantic Seaboard for shipment to Europe because the railroads were unable to respond quickly enough. Almost overnight, rail traffic began to seem obsolete. Even in the Niagara fruit belt, farmers were beginning to use trucks to haul their crops to market. Highways were suddenly the "in" thing. The infant trucking industry boomed and the radials were doomed.

Publicly funded highways were also attractive for political reasons. They could be used both as vote getters and as payoffs to corporate contributors, who could be awarded road construction or supply contracts. They also fell squarely under provincial jurisdiction, unlike the federally-regulated railroads. For these and other reasons, the battles over Beck's radials heated up and consumed prodigious amounts of political energy. A few lines were actually built, but early results were generally discouraging, not to mention expensive. The "build more roads" lobby clashed with the "build more radials" lobby from town to town and into the Legislature. Premier Drury tried to put an end to the controversy with another Royal Commission. This one overlapped the Gregory Commission, but was specifically mandated to study only the radials. Chaired by Justice Sutherland, the Commission investigated the faltering viability of electrified railways in the U.S. and Canada. The trend away from railroads was documented. The North American love affair with the car was just beginning. The Sutherland Report, tabled in mid-1922, concluded firmly that the radials were a bad idea. But the real stake through the heart of Beck's scheme was the January 1, 1923, plebiscite in Toronto, the hub of the system. A previous radials referendum in 1916 had passed handily. The approved bylaw granted Hydro control over four rights-of-way into Toronto along the waterfront, but nothing had happened since then because of the legislative wartime moratorium on radials construction. Beck's plans had grown considerably in the intervening years, however, and in 1922 he asked Toronto City Council for control over more rights-of-way, six in total. This caused a problem with the new Toronto Transit Commission. An outgrowth of the municipal radials movement, the TTC took public control of Toronto's transit system after the Mackenzie Syndicate's monopoly expired in 1921.

The TTC was unwilling to surrender control over all rail rights-of-way in the city. A compromise was proposed giving the TTC jurisdiction over two of the rights of way and Hydro control over the other four. Beck refused this offer, insisting that all radials should be completely controlled by Hydro on behalf of all participating municipalities. After a fierce seven-day debate at City Hall, Beck got his way. Council voted to give Hydro control over all six rights-of-way. Premier Drury, however, insisted on a new referendum vote on the grounds that the enlarged radials plan had not been approved by the ratepayers. In what must have been the most bitter electoral defeat of the very few Beck had known, Toronto voters rejected the new plan by 55 percent.

Beck soldiered on, of course, to the severe detriment of his reputation. Still, his general idea of a network of rail lines radiating from Toronto survives to this day in the form of the GO Transit system. GO Transit is a provincial government agency that operates commuter trains (diesel-powered, sorry, Sir Adam) to and from Oshawa, Hamilton, Bradford, Georgetown, Stou-ffville and many points between. Together with its supplementary fleet of buses, the GO "radials" keep tens of thousands of smog-generating cars off the already intolerably congested highways in and around Toronto. A single 10-car GO train carries the same number of people as 1,400 automobiles. A single GO bus replaces 50 autos. Without the GO system, Toronto would need the equivalent of three new Queen Elizabeth Ways, three new Don Valley Parkways as well as two million gas masks to keep all the extra pollution from killing more of us than it already does. Although it proved impractical on the scale he envisioned, perhaps Beck's idea wasn't so entirely off-the-wall after all.

Knowing he was dying, Sir Adam Beck rallied himself to fight American private power interests that were still trying to discredit Hydro. He also worried aloud that short-sighted Ontario politicians would eventually destroy his creation.
Hydro One Archives

One of the chief opponents of the radials, naturally, was Syndicate partner Sir Henry Mill Pellatt, a founder of the Toronto Electric Light Company and the Toronto Street Railway. Pellatt no doubt devoured every report of Sir Adam's public scolding, his old foe brought low. But the quintessential plutocrat of Casa Loma had himself been brought even lower. The Toronto Electric Light Company was long gone, a victim of the public power movement. The Toronto Street Railway monopoly was now lost. The Syndicate's other power interests were either sold or folded. A series of poor investments, perhaps in his eagerness to re-establish himself as the financial baron he once was, ate up his fortune. Pellatt owned his own home, but upkeep on the $3.5 million, 98-room castle in mid-Toronto was getting to be onerous. The servants were paid an average of only $550 per year, but there were 40 of them. Maintenance costs were over $80,000 a year. Coal for heating cost $15,000. While Sir Adam was being upbraided by editorialists, Sir Henry was being hounded by creditors and angry investors. By 1924, he was $1.7 million in debt and could no longer pay the taxes on Casa Loma. He surrendered the castle to the city and moved to a King City farm with his ailing wife, who died in early 1925. Pellatt lived in poverty and obscurity until he died in 1939, when thousands of Torontonians lined the streets to witness his funeral procession.

Looking at today's power industry, the parallels to Pellatt's time are striking, except that today's private power barons do not have public power to blame for their financial woes. The root of their problems, ironically, lies with the privatization and deregulation they so much wanted. Ken Lay and Andrew Fastow of Enron (and executives in many other energy companies) may have been lying and not playing by the rules, but a so-called deregulated marketplace made that cheating possible. British Energy shareholders were riding high for a while after the Conservative government of the United Kingdom privatized that publicly owned nuclear power company; now they are picking over the bones of near-bankruptcy thanks to deregulation (see chapter 12).

Adam Beck died at his home in London, Ontario on August 15, 1925, of the pernicious anemia that had been diagnosed in March of that year. For five months, he knew he was dying. He devoted much of that time to replying in depth to a U.S. publication, allegedly from the Smithsonian Institute, the American national museum in Washington, D.C. The booklet, *Niagara Falls: Its Power, Possibilities and Preservation* pronounced Ontario's public power system a complete failure and called Beck everything but a Bolshevik. Before Beck died, a U.S. government enquiry fingered the National Electric Light Association of America (NELA) as the instigator of the "study." NELA, a powerful lobby group for U.S. private power interests, had tried before to discredit public power by fraudulently attacking its most successful example. Given all his travails in the past four years, it was only fair that Beck lived to see his old American nemesis exposed once again. His greatest fear was not the Americans, however, but the politicians in Ontario who would, if allowed, use Hydro for transient political objectives. Among his last recorded words were these: "I had hoped to live to forge a band of iron around the Hydro to prevent its destruction by the politicians."

7

Public Power in the United States

Ontario Hydro's success gave inspiration to the U.S. public power movement and New Deal President Franklin D. Roosevelt.

The peculiar Canadian tradition of having to be recognized in the United States before being fully respected at home may have started with Ontario Hydro. From the early years of the twentieth century through the 1930s, Ontario Hydro gave inspiration to the public power movement in the U.S. and was the progenitor of some of our neighbour's largest and most successful public power systems. Indeed, Adam Beck himself was invited to testify before the Committee of Water Powers of the House of Representatives in 1917. Beck boasted, with facts and figures, how much less Ontario's power costs were under the public rather than private system. He explained in detail how the HEPC had become, in only seven years, the largest transmitter and distributor of electricity in the world, even before the Queenston-Chippawa project. And he reassured the Americans that this had all been accomplished without the need for higher taxes. His presentation was a hit with the Congress, which publicly issued his statement in a 70-page booklet.

For most of the first half of the last century, however, the U.S. took a very different path to electrical power development than Ontario, and later most other Canadian provinces. In the first three decades of the 1900s, the complaints of American power consumers were the same as those heard in Ontario: high rates, low reliability, lack of service to areas deemed unprofitable. There were successful municipal public power campaigns throughout the U.S., including Seattle, Cleveland, Phoenix and Los Angeles. There was even an entire state that went public: Nebraska, which remains a public power haven to this day. But the vast majority of Americans who used electricity at that time got it from private power developers, many of them the offspring of Thomas Edison's pioneering utility, Edison General Electric, which established utilities in several major American cities, mainly in the northern half of the country. Edison sold out to financier J.P. Morgan in 1892 and his name was dropped from the company, though many of the regional utilities it owned retain the esteemed inventor's brand to this day. The Association of Edison Illuminating Companies, the lobby group founded in 1885 by the man himself, is still around and still pumping out propaganda for the privately owned utility sector.

To fend off competition from public power, another private power lobby group, the National Electric Light Association (NELA), mounted an expensive and sustained anti-public power campaign that began very early in the century and continued in one form or another until the early 1930s. The campaign aimed to persuade the general public and state governments that regulated private monopolies would be more efficient than public power. Why toy with socialism, argued innumerable unsigned newspaper editorials distributed by a NELA front organization, when there was a perfectly sensible way of melding profit motive with public good? Numerous NELA-funded studies signed by hired academics were churned out, many finding their way into influential publications. Millions of pamphlets were delivered to opinion leaders around the country. Battalions of speakers were dispatched to business and service clubs, even churches. Private utilities paid to have chapters promoting the idea of private power monopolies included in school textbooks.

The NELA strategy of freezing out public power by agreeing to regulation in return for monopolies was largely successful. New York was first to establish an electric utility regulator, in 1907. By 1912, more than half of all 48 states had utility regulatory commissions, by 1934 there were 40. These commissions generally had the authority to franchise the utilities, regulate their rates and ensure that service was provided to everyone in the franchise area—the "obligation to serve" principle. Profits were regulated at levels considered "just and reasonable" but the profit-driven utilities naturally found them insufficient. A structural method of getting around regulation then evolved: the utility holding company. The holding company was a corporate entity that owned utilities in more than one state. The significance is that the U.S. Constitution gives the federal government all authority over interstate commerce, which means that if you operate in more than one state, you are subject to federal, not state laws. The holding companies were therefore not subject to state regulation and there was no federal regulation of utilities.

If the state regulators could, in the end, dictate rates and service standards, did it really matter who owned the local utility? It mattered a lot. The holding company could effortlessly siphon extra profits out of its subsidiaries in various ways. High management and other corporate service fees had to be paid to the parent company. Vertically integrated supply lines forced the local utilities to pay excessive prices to their parent holding company for their equipment and materials, or the holding company bought real estate that was resold to its utilities for wildly inflated sums. The local utility then appealed to the state regulator for higher rates to cover the increased cost of operations, plus, of course, a "just and reasonable" rate of return. That was another part of the scam. The capitalization on which that rate of return was based was usually inflated through phantom investments, a fact that later came out in an investigation by the U.S. Federal Trade Commission. In these ways, the local and regional utilities became enormous cash cows for the holding companies. State regulators could investigate these extortionate arrangements, newspapers could expose them and progressive politicians could fulminate against them—all of which happened frequently—but in the end, the higher rates were usually approved simply because there were no alternatives to the private monopolies. "Go ahead, deny us the higher profits" the private utilities would say. "But who will you call when the lights go out? As for the public power option, don't be stupid. All the studies on Ontario Hydro show that its allegedly cheaper rates are a fraud and that it is bringing the province to the point of fiscal collapse." So went the myth.

As the holding companies grew more profitable, they were able to buy more local utilities. By 1921, privately owned utilities were providing 94 percent of total generation, publicly owned utilities the remaining six percent. At their peak in the late 1920s, the 16 largest electric power holding companies controlled more than 75 percent of all U.S. generation. Of these 16, three dominated: The Southern Company, now known as North Carolina's Duke Energy, (after its founder, James Duke); The Morgan Group, which now includes companies such as New York's Consolidated Edison and Public Service of New Jersey, and Samuel Insull's Middle West Utilities, which still survives as Commonwealth Edison of Illinois, which itself is a subsidiary of Philadelphia-based mega-utility Exelon, the largest nuclear operator in North America. Of these three, Insull was generally regarded as America's chief power baron. And as NELA's president, he was chief strategist for the private monopoly and anti-public power campaigns. His rise and fall was uncannily like that of Ontario's own private power baron, Sir Henry Pellatt.

Samuel Insull's private utility "holding company" empire collapsed in 1932; tens of thousands of shareholders lost billions of dollars. It is widely believed that Insull was the model for the trademark plutocrat of the Monopoly board game, which was invented in 1930.

Samuel Insull was born in London, England, in 1859, the same year Pellatt was born in Kingston, Ontario. An admirer of Thomas Edison, Insull went to America in 1881 to be Edison's personal secretary and quickly worked his way up to vice president of the Edison General Electric Company. When J.P. Morgan bought General Electric in 1892, Insull was sent west to Chicago to run the struggling Chicago Edison Company. Aggressive and clever, Insull bought out all of his competitors for peanuts after the Panic of 1893 and by 1907, all of Chicago's electricity was being generated by Insull's Commonwealth Edison Company. Like Pellatt, he also controlled the city's transit system. Insull not only bought out his rivals, he also bought the exclusive rights to all electrical generating equipment from every U.S. manufacturer, including the largest, General Electric. It was pretty tough for his potential competitors to get into the business when they couldn't buy the necessary equipment.

By 1912, Insull had created Middle West Utilities and other holding companies that would eventually control well over 12 percent of the entire U.S. power market. His companies operated over 600 generating stations in 30 states and his political power was just as expansive. It was mainly Insull that U.S. New Deal-era Senator George Norris, the political father of the Tennessee Valley Authority, was referring to when he said, "the power trust was the greatest monopolistic corporation that has been organized for private greed.... It has bought and sold legislatures. It has interested itself in the election of public officials, from school directors to the President of the United States." (And Norris was a Republican!) Indeed, Insull made generous contributions to the Republican Party and had ready access to the very highest officials in the U.S.

Insull seemed invincible and as the power industry expanded, bankers lined up to loan him as much money as he wanted. As his stock soared in value, so did his indebtedness, secured with the stock. But like Pellatt before him, Insull lost it all. In 1932, three years after the Wall Street crash of 1929, Insull's empire also crashed, along with his political influence. Tens of thousands of people lost their life savings in the meltdown. Losses were estimated as high as $2 billion, a foreshadowing of the Enron collapse seven decades later. In disgrace, Insull surrendered his holding companies to his creditors and, facing indictment for embezzlement and fraud, fled the country with as much cash as he could stuff in his suitcase. He went to Greece and then Turkey, which did not have an extradition treaty with the U.S. So great was the public outcry, however, that he was eventually returned to the U.S. to face the music. To the amazement and anger of the country, he was acquitted of all charges and once again left the U.S. He died of a heart attack in a Paris subway station in 1938, a year before Pellatt. Like Pellatt, he was penniless at the end. Unlike Pellatt, he was not given a grand funeral.

Insull's holding companies were not the only ones to implode in the years after the Wall Street disaster. Despite the widespread financial calamity, Republican President Herbert Hoover continued to maintain there was no reason to impose federal regulation on the power industry. No doubt Insull's influence extended to the upper reaches of the Administration (as did that of Enron Chairman Kenneth Lay, the now-disgraced energy policy advisor and major campaign contributor to current U.S. President George W. Bush).

When Franklin D. Roosevelt and the Democrats swept the 1932 election, correcting the abuses of private power became a government priority. This led to the Public Utility Holding Company Act of 1935 (PUHCA), which put holding companies under the regulation of the Securities and Exchange Commission (SEC). As well, utilities involved in interstate wholesale marketing or transmission of electric power became regulated by the Federal Power Commission, which evolved into the Federal Energy Regulatory Commission (FERC).

Insull became emblematic of the corruption and fraud that contributed to the Great Depression. Invoking his name alone made it easy for the Roosevelt Administration to roll over private power's objections to the creation of federally owned utilities such as the Tennessee Valley Authority (TVA) and the Bonneville Power Administration, as well as the Rural Electrification Administration. U.S. government engineers were frequent visitors to the Sir Adam Beck station at Queenston, learning what they could to help them build Nevada's ironically named

Hoover Dam and Washington State's gargantuan Grand Coulee Dam, of which Americans are justly proud. Don't bet on these monuments to American public enterprise being privatized any time soon.

From 1933 to 1941, one half of all new generating capacity in the U.S. was provided by federal and other public power agencies. By the end of Roosevelt's second term, in 1941, public power contributed 12 percent of total U.S. utility generation, double what it had been 20 years earlier. That number has doubled again in the ensuing years; 25 percent of all power in the U.S. is now public. In Canada it is approximately 80 percent.

In the late 1920s, the truth about Ontario Hydro finally began to emerge from the thicket of lies and distortions funded, in large part, by American private power interests. In 1927, a U.S. Senate committee began hearing testimony about private power's astoundingly well-financed and pervasive campaign to systematically discredit public power, using means most Americans found offensive. One of the first known American academic studies of Ontario Hydro conducted without private utility funding was published in 1929 by New York's Syracuse University. The study compared Ontario Hydro's rates and operations to those of private power companies then operating in New York State. It found that the oft-repeated contention by previous studies—that Ontario industrial customers were subsidizing small users—was a myth. It also praised the managerial efficiency of the HEPC:

> The Ontario system has demonstrated the feasibility of public ownership and operation of this utility on a thoroughly businesslike basis and on a statewide scale. It demonstrates that it is entirely possible for the government to secure the services of competent managers and technicians if it is willing to pay adequate salaries and make appointments on merit. It is far from axiomatic that government enterprise is necessarily and inevitably inefficient...
>
> It is not inconceivable that the Ontario experiment may serve as a beacon light to the people of the United States, should a wave of protest ever get underway at the methods of those in control of the industry on this side of the Canadian border.

The wave of protest was already rising. Even before this study, the American public power movement was fascinated with the Ontario experiment. One of the most influential organizations in that movement was the Public Ownership League, founded in 1916 by Wisconsin

Senator Robert LaFollette, a Progressive, and Wisconsin state legislator Carl D. Thompson. The basic idea of the League was essentially a clone of the Ontario system: public utilities, which are natural and essential monopolies, should be publicly owned and developed. Thompson would devote the next 33 years of his life to public power, frequently citing the Ontario model as an example of its success. Ever on the alert for such godless socialists, NELA tracked Thompson religiously and hounded him incessantly. His meetings were disrupted by private power operatives; propagandists were sent in his wake to counter his impact. In 1924, Thompson organized a National Public Ownership Conference with support from the American Federation of Labor, the Electrical Union, the National Grange and the League for Industrial Democracy. The conference proposed a nationwide public-power system that looked remarkably like Ontario's and a bill to that effect was introduced the next year into Congress. That proposal was the forerunner of what became the Roosevelt-era push for public power.

Franklin Delano Roosevelt. As governor of New York (1929-32), Roosevelt established the New York Power Authority, which closely resembled Ontario Hydro. As U.S. President (1933-45), he created the Tennessee Valley Authority, the Bonneville Power Administration and the Rural Electrification Administration.

As Governor of New York in the four years before he was elected President, Roosevelt became enamoured of Ontario Hydro's achievements. But he wasn't the first New York governor to notice what was happening on the other side of the border. In 1907, Governor Charles Evans Hughes (later Chief Justice of the U.S. Supreme Court) said that water power in his state "should be preserved and held for the benefit of the people and should not be surrendered to private interests." Possibly Hughes was inspired by Ontario Premier Whitney's 1905 declaration that our province's water power "should not in the future be made the sport and prey of capitalists and shall not be treated as anything else but a valuable asset of the people of Ontario." In 1914, Theodore Roosevelt, a cousin of Franklin's who had been New York governor before Hughes, as well as U.S. President, warned against the "waterpower barons" seeking a monopoly on New York's natural resources. And throughout much of the 1920s, Governor Alfred E. Smith tried to emulate Ontario by developing hydroelectric power through a state agency. He was continuously thwarted by private power interests.

In his inaugural speech as governor in January 1929, Roosevelt said, echoing Whitney and Hughes, that "the water power of the State should belong to all the people. The title to this power must rest forever in the people." He was attacked virulently by the utility holding companies, but made it his solemn mission, which he often reiterated, "to give back to the people the waterpower which is theirs." In his first year in office, Roosevelt established a commission to study the hydroelectric potential of the St. Lawrence River. It is very likely that the Commission saw the Syracuse University study noted earlier, as many of its recommendations reflected confidence in the basic model of public power in Ontario. On April 27, 1931, Roosevelt signed the bill that established the New York Power Authority. The NYPA was then, and still is, America's largest state-owned public power enterprise. It sells power at cost to government agencies and to over 50 municipal or cooperative-owned distribution systems. It also provides affordable power to job-creating industries in the state.

One year after signing what he called the most important legislation of his governorship, Roosevelt, this time as a presidential candidate, made a magnificent speech on public power in Portland, Oregon. Had he been able to hear it, Sir Adam Beck would have risen from his grave to applaud. In it, Roosevelt proclaimed his belief that "the question of power, of electrical development and distribution, is primarily a national problem." He roundly condemned the "systematic, subtle, deliberate and unprincipled campaign of misinformation, of propaganda, and, if I may use the words, of lies and falsehoods… bought and paid for by certain private utility corporations." He traced the idea of public authority over waterways back to extortionate ferry-boat operators in the time of King James, and said that the regulatory commissions of many states "have often failed to live up to the very high purpose for which they were created." He spoke favourably of Canada, where "the average home uses twice as much electric power per family as we do in the United States" and said the reason for this was "frankly and definitely that many selfish interests in control of light and power industries have not been sufficiently far-sighted to establish rates low enough to encourage widespread public use." Roosevelt then laid out his vision of the future of federal government power policy, which included the right of municipalities, through the authority derived from referendum votes, to establish publicly owned utilities. He ended with a reprise of his 1929 inaugural speech: "As an important part of this policy the natural hydroelectric power resources belonging to the people of the United States, or the several States, shall remain forever in their possession." Once he became President, Roosevelt acted quickly on that promise, as he had in New York.

It is fitting indeed that New York's largest hydropower development is known as the Franklin D. Roosevelt Power Project. It is even more fitting that it is adjacent to Ontario's second-largest hydroelectric project, the 1,000 megawatt Robert H. Saunders Generating Station, near Cornwall. The Roosevelt project, a massive, 60 kilometre-long channelling of water from the St. Lawrence River as it rushes headlong towards the Atlantic Ocean, feeds New York's Robert Moses Generating Station, which also produces about 1,000 megawatts of

Queen Elizabeth II chats with U.S. Vice President Richard M. Nixon following the 1959 unveiling of the inscription on the monument at the midpoint of the joint New York-Ontario hydroelectric power project on the St. Lawrence River, near Cornwall. Sir Adam Beck had advocated such a project to the U.S. Congress 42 years earlier. *Hydro One Archives*

emission-free power. Robert Moses meets Robert Saunders at the U.S./Canada boundary in the middle of the waterway. An 18-metre silvery arch straddles the boundary, one foot in each country. Under the arch is a large black granite monument that was unveiled by Queen Elizabeth II on June 27, 1959. Engraved in the granite are the coats of arms of Canada and the United States, along with these words:

> This stone bears witness to the common purpose of two nations, whose frontiers are the frontiers of friendship, whose ways are the ways of freedom, and whose works are the works of peace.

Some may cynically remark: "If only that were so," including myself, at times. But the values these words honour—friendship, freedom, peace—are surely the compass points that will lead both our societies to a better future, if only we follow them. And there can be no better place to permanently enshrine our mutual expression of these ideals than at the meeting place of two of our continent's greatest examples of the power of public purpose.

8
The Politics of Power

In pre-World War II Ontario, Hydro's struggles to keep up with rapidly growing demand for power were plagued by political interference and manufactured controversy.

For many years, Ontario Hydro has been widely criticized for its edifice complex. And justly so. Until it was abruptly stopped in its tracks ten years ago by the NDP government of Premier Bob Rae, Hydro's long-standing plan for more generating stations, mostly nuclear and coal, would have burdened Ontario with well over twice the power capacity it now needs on its hottest or coldest days. The impact on our hydro rates would have been intolerable, the impact on our environment horrific. In both the NDP caucus and Cabinet during the government of 1990-95, discussions on what to do about Hydro were often the most intense.

The questions that were asked in 1990 at the Cabinet table are the questions that need to be asked again in the current public debate about our energy future. How did Ontario Hydro become the captive of a narrow and single-minded construction mania that saw the energy future strictly in terms of more nuclear plants and greater use of dirty coal? Why were energy conservation strategies never considered? How and why did the nuclear construction mania become embedded in Ontario Hydro's culture? While the answers to these questions are far-reaching and complex, the point of entry is a basic power supply problem.

Put yourself in the position of the public power system planner. Lacking a crystal ball, you still have to peer many years, often decades, into the future and try to anticipate the need for electricity. Generating stations and transmission lines are very expensive and take years to build. Paying off their construction and financing costs must be spread out over 30 to 40 or more years. The cost of these assets is "sunk" because there's no other economic use for transmission towers, transformer stations and power plants. If you don't build enough capacity to meet growing demand, you're the idiot whose short-sightedness is hampering the province's economic development. If you build too much capacity that is idle most or all of the time, you're the idiot who has burdened hydro ratepayers with the high rates needed to pay for the capacity even though it is not being used. What do you do?

Most of us would undoubtedly want to err on the side of building too much capacity rather than too little. A "depression mentality" of having a little extra put away in case of bad times is easily sustained in the public power business. For customers, carrying the cost of too much capacity, while bothersome, is much less than the cost, insecurity and disruption of having too little power. If your manufacturing costs rise because of higher hydro rates, that's a drag. If you can't manufacture anything because you have no power, that's a disaster. Not just for you, but for your workers, their families and communities, and for the rest of us who must carry the heavier public and private burden that inevitably follows reduced economic activity. The outcomes are reduced employment, reduced public services, higher public debt or some combination of all three. With all of this in mind, the complex and often-controversial history of Ontario Hydro post-Adam Beck becomes easier to fathom, and the failure of a banal succession of Conservative and Liberal governments to integrate power system planning with economic (and later, environmental) policy becomes more obvious.

Prior to the mid-1950s, when the entire province was finally electrified, Ontario Hydro did need to build, build, build in order to meet demand growth that more than doubled every decade. Except for a three-year period in the early years of the Great Depression, Hydro had always been in a race to find and deliver enough power to feed a rapidly growing industrial economy that had come to utterly depend on reliable, affordable hydroelectricity. And it had to be hydroelectricity; dependence on American coal was out of the question. Finally, the closer the generation was to the population centres of southern Ontario, the more affordable it would be.

After Niagara had been tapped to the limits allowed by the 1909 Boundary Water Treaty with the U.S., the most obvious hydropower resource for southern Ontario was the mighty St. Lawrence River, the last conduit through which the water of all the Great Lakes ultimately flows downhill to the Atlantic. The St. Lawrence begins at the eastern edge of Lake Ontario, near Kingston, Ontario's first capital. From there until Cornwall, very near the Ontario-Quebec boundary, the middle of

Ontario Hydro Chairman Robert H. Saunders (right) joins U.S. President Harry S. Truman and Michigan Governor G. Mennen Williams in Detroit in 1951 for the final stages of lengthy power negotiations between Canada and the U.S. that paved the way for the St. Lawrence project. Four years later, Saunders died from injuries suffered in a plane crash near London, Ontario. The Ontario portion of the St. Lawrence project was named after him. *Hydro One Archives*

the river defines the border between Canada and the U.S., specifically New York State. At that point, because of lines drawn on the map at the 1783 Treaty of Paris that ended the American Revolutionary War, the river ceases being an international waterway and passes entirely through southern Quebec, making it a provincial waterway.

As early as 1913, Adam Beck was dreaming of power development on the St. Lawrence as part of a grand scheme to link the interior of the continent with the Atlantic by making the entire river navigable to ocean-going ships. Before Congress in his 1917 visit to Washington, Beck had urged American legislators to cooperate with Canada in developing the international waterway. Back in Toronto, he predicted that the St. Lawrence project was imminent. In fact the idea of such a seaway was already centuries old by the time Beck came along. The Niagara Region's Welland Canal, which makes navigation between Lakes Erie and Ontario possible, was built in 1829, and later expanded. The joint U.S.-Canadian Deep Waterways Commission was formed in 1895 to study the feasibility of a seaway. In 1909, under the Water Boundary Treaty, this Commission was entrenched as the International Joint Commission (IJC), which to this day has jurisdiction over U.S.-

Canadian boundary waters.

While the structures and institutions were in place, politics on both sides of the river delayed seaway and power developments in the international section of the St. Lawrence for nearly half a century, until 1951. In Quebec, however, there was no need to deal with the Americans on these matters. Indeed, power projects on the river near Montreal had pre-dated Ontario Hydro's development on the Niagara River.

Charles A. Magrath, Adam Beck's successor as Chairman of the Hydro-Electric Power Commission of Ontario, was Chairman of the Canadian section of the IJC between 1914 and 1935. An extraordinarily competent and politically astute man, and as devoted to public power as Beck was, Magrath knew that it would be many years before Ontario would be able to tap its side of the St. Lawrence for power. The political combination of the heavily populated U.S. port cities on the Atlantic seaboard, the railroad interests that moved all the freight inland and the private power interests that wanted the St. Lawrence for their own, proved too difficult for even U.S. Presidents like Franklin Roosevelt to overcome. It wasn't until after World War II that public power projects on the St. Lawrence were pursued.

The second most obvious hydropower resource was the Ottawa River, which serves as the Ontario-Quebec boundary from Haileybury in the north to just below Hawkesbury in the south, where it meets the St. Lawrence, near Montreal. Here again, Magrath was frustrated by politics. Hydroelectric development on the Ottawa River was complicated by the need for an agreement between the Ontario and Quebec governments. This was unlikely to happen while Quebec had a large surplus of private power, which was the case in the mid-1920s (public power in Quebec was still some 20 years away).

When Magrath took over Hydro in 1925, Ontario was again coming perilously close to an acute power shortage. Private power companies were being bought out by the HEPC, but this added no new capacity. Northern rivers had a lot of hydroelectric potential that would eventually be tapped (some of it already had been by private developers, mostly mining and forest product companies), but the cost of building transmission lines to the south was prohibitive at that time. Magrath considered building coal-fired steam generating plants, but rejected the idea, thankfully. He felt he had no choice but to turn to private power developers inside Quebec, with their abundant and relatively inexpensive power. Since the HEPC had been purchasing some private power from 1910 on, this was hardly a precedent. But the province's pride in its accomplishments at Niagara and the by now near-unanimous endorsement of public power in Ontario made it necessary for Magrath to explain that buying private power from Quebec was simply an interim measure until more Ontario sites could be built. Premier Ferguson's Conservative government agreed and power purchase contracts with Quebec companies were quickly negotiated.

The first contract, signed in 1926, was with the Gatineau Power Company, which was already building power generating stations on the Gatineau River near Ottawa. Magrath planned to bring the power from there to Toronto over a 370 kilometre transmission line that would have to be built over largely uninhabited terrain. The line would then terminate at the yet-to-be-constructed Leaside Transformer Station where the new power could be fed into the Niagara transmission grid. Both the line and the transformer station were, for that time, great feats of power system engineering and further evidence that Ontario's public employees, as the Syracuse University study of 1929 concluded, were as industrious and innovative as any in the private sector. The line was longer than any that had yet been built in Canada and carried 230,000 volts, an unprecedented amount of power. In one of the world's first uses of aerial surveying, the path of the power lines was plotted from three kilometres up and its towers sited with stereoscopic cameras. Completed in only one year, the Gatineau line was so well built that it was virtually problem-free for the next 70 years, until the devastating ice storm of January 1998 brought down many of its towers.

After completing the Gatineau line in 1928, the Hydro crews began constructing an even longer (530 km) line from the Beauharnois Light, Heat and Power Company plant on the St. Lawrence near Montreal. It too would terminate at Toronto's Leaside Station. Magrath felt that the Quebec contracts would give the HEPC some breathing room to plan and build new facilities in Ontario as economically and as thoughtfully as possible. He was right, but for a few politically tumultuous years, he was *too* right.

In 1931, because of the Depression, demand for power in Ontario dropped for the first time since the HEPC had begun delivering it 21 years earlier. As the economy bottomed out, so did power usage. The electricity from Quebec was not needed, but had to be paid for anyway. Moreover, the HEPC was at the same time selling much cheaper power from Niagara to the U.S., an obligation it had contractually inherited when it bought out the Ontario Power Company in 1917. As these facts became more widely known, criticism of Hydro began to mount and spilled over into the Legislature. The Liberals, under their 35-year-old leader, Mitchell F. Hepburn, found in the Quebec contract and Niagara export controversies issues with which they could pillory the Conservatives, now led by Ferguson's successor, George S. Henry.

Hepburn was only 30 years old in 1926 when he was elected in the London-area riding of Elgin West as the youngest-ever member of the House of Commons. In 1930, while still in Parliament, he was elected Ontario Liberal Leader and soon thereafter began attacking Hydro as an irresponsible and spendthrift agency of an old and corrupt Conservative government. Since the Liberals had lost their decades-long rule of Ontario in 1905 largely because of their opposition to public power, the party understandably had no philosophical or political commitment to Hydro. This was likely exacerbated, in Hepburn's case, by an injustice he felt he had suffered at age 16 in an incident that involved Adam Beck.

In October 1912, Beck was personally conducting a demonstration of electrical farming appliances, including an electric cow milker, to a large crowd of 3,000 people near St. Thomas, where Hepburn was attending high school. Along with a couple of school friends, Hepburn climbed up a windmill close to the action in order to get a better look. Beck, who was wearing his trademark black bowler, stepped onto a platform to address the crowd. A single apple thrown from the windmill knocked off Beck's hat, probably eliciting laughter from the crowd and most certainly from the fleeing teenagers. When he returned to school two days later, Hepburn was accused of throwing the apple. His vigorous denials were not believed and he was ordered to leave school until he personally apologized to Adam Beck or revealed who had actually thrown the apple. Outraged, Hepburn stormed back to his history class, scooped up all his books from his desk, snatched his cap from its peg, and walked out the door without a word to anyone. He never returned to school.

In 1931, Hepburn launched an all-out attack on Hydro, claiming it had been corrupted by politics. Rates are too high, he repeatedly said, and he demanded an investigation into the Beauharnois contract, an investigation the government agreed to by setting up a Special Committee of Inquiry. A few months later, Hepburn claimed that the Inquiry was not enough; a Royal Commission was called for, not just to look into the Beauharnois contract, but into all recent contracts and projects of Ontario Hydro. Styling himself a champion of the downtrodden victims of government economic mismanagement, Hepburn planted, fertilized and harvested anti-Hydro sentiments everywhere he went. He promised to clean Hydro's house if elected Premier and vowed to cancel the Quebec contracts. No amount of explanation by the HEPC, now headed by another Beck disciple, J. R. Cooke, would satisfy the Liberals, who continued to demand exhaustive investigations.

Though he claimed to be a great supporter of public ownership,

Hepburn's relentless attacks on Hydro seemed to be in the tradition of the anti-public power propagandists of years past. In fact, they were. In February 1932, Premier Henry read to the Legislature passages from a standard-issue anti-public power pamphlet published in Chicago (the centre of Samuel Insull's empire, which was about to crash). Henry then matched these passages word for word with excerpts from speeches given by Hepburn. Henry scathingly called Hepburn someone who "stands up and expresses faith in Hydro and public ownership, and then seizes a copy of an organ directly built for the purpose of destroying public ownership anywhere."

It was an embarrassing but temporary setback for Hepburn, who continued to harp on Hydro's alleged mismanagement and corruption. Later that year, Premier Henry finally appointed a Royal Commission to investigate serious accusations of Hydro-related financial improprieties. One of the allegations was that a certain John J. Aird, who had a consulting relationship with Hydro, had received a payment of $125,000 from the Beauharnois Corporation after its contract with Hydro had been signed. The Royal Commission not only confirmed this payment, but heard from R.O. Sweezey, president of Beauharnois, that he had given the money to Aird for the Ontario Conservative Party. Aird categorically denied this, claiming that the money was paid strictly for consulting services rendered; no political party had received a nickel of it. The Royal Commission could find no evidence to the contrary. In deciding whether Aird or Sweezey was telling the truth, it came down on the side of Aird, concluding that his payment had no connection with Hydro's Beauharnois contract.

It is now 70 years later and I have not seen any of the evidence, but this finding seems suspicious. $125,000 at that time translates into well over a million dollars today. That's a lot of money for what must have been a straightforward "consulting" assignment. And what did Sweezey have to gain by accusing the party in power in Ontario of taking payoffs to ensure a contract that enriched Beauharnois shareholders? After all, Hepburn had promised to cancel the Quebec contracts if elected. Why hand him an indisputably sound reason for doing so? Either we're missing a piece of the puzzle that truly exonerates Aird and the Conservatives or we're dealing with a rather pliant Royal Commission. Personally, I hope the former explanation is correct. I would be distressed to know that a Royal Commission, even one from seven decades in the past, would willingly overlook a patently illegitimate "public-private partnership."

If the Royal Commission's findings do not satisfy me, you can imagine how little they affected Hepburn, whose attacks on the Quebec contracts went on undiminished. In 1933, a new issue arose surrounding the province's 1932 purchase of privately owned Ontario Power Service Corporation's uncompleted Abitibi Canyon power development. Premier Henry had personally conducted the negotiations. In a plot reminiscent of the 1907 exposure of Liberal Premier Ross's financial interest through an insurance company in the privately owned Electrical Power Development Company, it came to light that Henry himself had a $25,000 stake in the Abitibi Canyon project. Moreover, he was a director of an insurance company that held a $200,000 interest in the same project. Henry's explanation of the apparent conflict of interest was weak and unconvincing. A later Royal Commission said flatly that he should not have participated in the Abitibi negotiations.

The 1934 election campaign was as brutal as any the province had seen. Henry (ironically) accused the Liberals of taking payoffs from Beauharnois. Hepburn pounded away at Ontario Hydro. The HEPC leadership, distressed at the years of negative publicity and fearing a Liberal victory, stupidly got involved in the election by publishing a pamphlet entitled: *Paid-for Propaganda*. The pamphlet accused the Liberals of adopting the dishonest tactics of the National Electric Light Association. Even worse, HEPC funds were spent on detectives who tried to find out if the Liberals were taking money from private power interests in the U.S. or Canada, as Henry had charged. Hydro's involvement in the election was completely wrong, which made it even easier for Hepburn to fire Chairman Cooke and practically everyone else in sight at Hydro after he won the June 8 election. Virtually all senior Hydro officers were replaced with Liberal appointees. The new Chairman was T. Stewart Lyon, editor of *The Globe*, at that time a Liberal newspaper. I can't help wondering if, after ordering the most politically motivated purge in Hydro's history, Hepburn ever stopped by Sir Adam Beck's grave to say: "I paid you back."

Hepburn made good on his promise to cancel the Quebec contracts, a move that temporarily caused panic in the bond markets. For a while, no one would loan money to the Province. Hydro was ordered to stop taking power from Quebec, and did so. Hepburn also intended to privatize all Ontario Hydro construction, but was talked out of it by the newly appointed Chief Engineer, Thomas Hogg, who would become Chairman in 1937. (Years later, in 1948, Hogg would become the second Ontario Hydro Chairman to be politically dismissed, this time by a Conservative premier, George Drew. The reason: power shortages. Hogg had miscalculated what the post-war supply/demand balance would be.) It was soon apparent, however, that Magrath had been right and Hepburn had been wrong. In 1934, Hydro demand began to rap-

The eerily Soviet-style deification of Sir Adam Beck at the 1936 Canadian National Exhibition is perhaps partially explained by the bitter conflict between Ontario Hydro management and recently elected Liberal Premier Mitchell Hepburn. Hepburn had turned Hydro into a political football during the Depression, cancelled projects and power contracts and made an attempt to partially privatize construction. Hydro management no doubt thought the ghost of Beck had more credibility with the public than the notoriously combative Hepburn. *Hydro One Archives*

idly recover. Ontario would very soon not be able to produce enough power to meet its own needs. Some of the Quebec contracts were discreetly renegotiated, albeit for lesser volumes at lower prices. Some of the private companies, including Beauharnois, were able to get court judgments validating their original contracts, but they too, in the end, decided to renegotiate. As well, many of the Hydro projects that were cancelled in the first flush of the Liberal victory were resumed over the next couple of years. Still, the province struggled with the constant threat of shortages during the Hepburn years, and rationing was imposed at the outbreak of World War II, in September 1939. Fortunately for the northern parts of the province, Ontario Hydro had built and bought enough generation capacity to sustain the long period of growth in resource industries that began in the 1920s and continued into the 1980s.

Hepburn was re-elected in 1937, but his reputation was ultimately doomed by his simplistic and uncompromising approach to complex issues, his contempt for people on relief during the Depression, his inability to get along with Liberal Prime Minister Mackenzie King (who once publicly called Hepburn a fascist) and his pathological hatred of industrial unions. He resigned as Premier and Liberal Leader in 1942, with no protest from his fractured party. In the 1943 election, the Conservatives formed a minority government with 38 seats. The Co-operative Commonwealth Federation, forerunner of the New Democrats, won 34 seats. Hepburn, who made a career of playing political football with Ontario's power system, saw his Liberals reduced to 15 seats. They wouldn't improve on that for another five elections.

You may wonder why such a strong believer in public power is recounting episodes of political interference and possible corruption in Ontario Hydro's history. Doesn't all this show that public power, because it can be so easily politicized, is inherently less stable and less focused on the job at hand than private power would be? The answer is no, it doesn't show that at all. Certainly if we're talking about stability, the current situation in the global utility industry provides no evidence that the private sector is at all better. The opposite is true, you only have to read the business press. Nor have private utilities and energy marketers focused on the job at hand: providing their customers with affordable, reliable power and their shareholders with steady, reasonable dividends. This is why, for example, there is growing momentum in the direction of more public power in that haven of deregulated private enterprise, the United States. (see Chapter 18).

I'm not saying that public enterprise is impervious to poor judgment or wrongful political influence on the part of individuals. People are people and some will be corruptible, though a private, less publicly accountable system makes corruption more likely. Most of us, however, wherever we work, try to do a good job, are honest and reliable and want to be respected for our efforts and loyalty. But the rules we work under naturally affect our personal expression of these qualities. Which takes us back to the original issue of how Ontario Hydro acquired its edifice complex.

Ontario Hydro grew up and spent most of its life facing the twin fears of power shortages and dependence on external suppliers whose prices could not be controlled, only negotiated. It was only natural that the Hydro culture would evolve towards monumentalism, especially since, beginning with Queenston-Chippawa, it seemed to work. The people of the post-war Hydro who planned all those coal and nuclear stations that are giving us so much trouble today, were most definitely doing what they saw to be their job. They *had* to be thinking 20, 30 and more years down the road and they had to rely on their own experience, or rather the experience of those who came before them. What they lacked was a government—a shareholder-in-chief—that had a vision of our province's energy future; a government that saw progress as using proportionately *less* energy, and more environmentally benign energy to achieve the same goals of Adam Beck and his contemporaries: competitive industries, productive farms, well-lit streets and comfortable homes.

And lest any diehard Conservatives reading this are tempted to protest: "Hey, our vision of Ontario's electricity system is different from the past!" I would remind them that their view of how things should work dates back to the unregulated private power system of a century ago. Not a different vision, just an older one. The people of Ontario voted against such a vision back then and would, if the Conservatives had the courage to put it to them directly, vote against it now.

"In Ontario, Hydro and Progress Go Hand in Hand." In 1957, a decorated Hydro truck helps the St. Lawrence village of Iroquois celebrate its Centennial. A few months later, along with several other villages near Morrisburg, Iroquois residents were relocated a few kilometres away before their land was flooded as part of the St. Lawrence power project. For many, unhappiness at being uprooted lingered for decades. Northern First Nations communities have had similar Hydro intrusion experiences. Ensuring that all legitimate past grievances related to Hydro operations are fully resolved is an important government responsibility. *Hydro One Archives*

9

The Natural Gas Betrayal

Federal Liberal and provincial Conservative decisions to give the young natural gas industry to private interests touched off an energy "civil war" and led to Hydro mistakes that still burden Ontario.

The 1950s were years of great hydropower accomplishments for Ontario Hydro. In 1954, the Dowager Duchess of Kent was on hand to help open the Sir Adam Beck 2 station, which had been made possible by the Canada-U.S. Niagara Diversion Treaty of 1950. In the Northwest, the Caribou Falls, Manitou Falls, Silver Falls, Whitedog Falls and Pine Portage stations began operating in this decade. On the Ottawa-St. Lawrence System, Hydro completed the Chenaux, Des Joachims, Otto Holden and the Robert H. Saunders which was opened by Queen Elizabeth II. Fifty years later, these stations collectively continue to produce more than 3,300 megawatts of clean, renewable and ultra low-cost power, about 20 percent of Ontario's average daily demand.

The 1950s also saw the construction of thousands of kilometres of new transmission lines and tens of thousands of kilometres of rural distribution lines. Today, the combined transmission and distribution lines of Hydro One (one of Ontario Hydro's successor companies), could circle our planet more than four times.

If you look at all the hydroelectric and network construction of the 1950s, those years could be justifiably labelled Ontario Hydro's finest decade. But two momentous political decisions during that decade—one in Ottawa, the other at Queen's Park—touched off a "civil war" in the Canadian energy sector that led to the horrendous coal and nuclear mistakes of the following three decades, from the 1960s to the 1990s. We will continue to live with the effects of the coal mistakes for decades to come. Some of us will die before our time, victims of coal-generated air pollution. In the case of the nuclear mistakes, we can only hope that our descendants can manage, without incident, the leftover radioactive shells of our decommissioned nuclear stations and the growing inventory of high-level radioactive wastes. Here in the present, the outrageous nuclear debt, and the increased hydro rates needed to retire it, have made it easier for first the Harris and now the Eves Conservatives, with the support of the Liberals, to hatch their privatization and deregulation scheme. The Liberals are now trying hard to give the impression that they're against privatization and deregulation because it's obviously not working as advertised, but the voting record is there for all to see. Moreover, the Liberals are almost as responsible for the nuclear debt as prior Conservative governments.

By 1950, the idea of building a natural gas pipeline from Western to Eastern Canada seemed imminently achievable. There was gas in great abundance in the West, mainly in Alberta, and it was easy to pump out of the ground. There were ready markets in the rapidly growing postwar economy of Ontario, as hungry as ever for affordable energy. Quebec would also be a major gas customer. The only issue was how to most cost-efficiently get the gas from there to here. To many people, the answer was so obvious it seemed like a wide-open net: build a publicly owned transmission pipeline that was either federally owned or jointly owned by the provinces through which it passed. Trunks off the main pipeline would eventually reach into every major population centre, the same as power transmission lines. Publicly owned local distribution systems—the underground networks of gas pipes we have to look out for when digging—would be built or upgraded, at first in the larger towns and cities and then into rural areas. This was exactly how the electrical distribution system gradually spread across Ontario from 1910 to 1960. The analogy between gas and electricity went even further. Since the cost of the fuel was low compared to the cost of delivery, just as with hydroelectricity, what mattered most to natural gas consumers was how much it cost to finance construction of the delivery system. And just as with electricity, government-backed entities would get much more favourable financing rates than the private sector. Calculations made in the early 1950s by Stanford University put the eventual cost of moving Western natural gas to Ontario consumers, if delivered through a privately owned pipeline, at 50-55 cents per thousand cubic feet. But most of that price was taken up with the financing costs of building the pipeline and, of course, the ever-necessary profits. A publicly financed, non-profit pipeline would be able to deliver the gas for 30 to 40 percent less.

It was the public vs. private power contest all over again, with another form of energy. This time, however, there was no reason for a contest. There was no entrenched private natural gas distribution industry in Ontario, as there had been with electricity. There were a few municipal gas distribution networks in the larger cities, but these had typically been built in the nineteenth century to distribute manufactured gas. These systems were usually primitive and unsuited to the objective of universal natural gas distribution. Indeed, with the advance of electricity, manufactured gas had become marginalized. By the 1950s "gaslight" was an archaic term used by older folks. The new natural gas industry was a completely open field. There was no one to buy out. No private investors could have cried foul if the government had simply moved in and developed it.

In Ontario, there was every reason to do just this. The province needed energy and for space heating and water heating, gas would be a lot cheaper than the then-dominant heating fuels of coal and oil, most of which was imported from the U.S. Certainly the ability of a public power system to reliably and more equitably deliver electrical energy at much lower average cost than private power had been established for over 40 years. There was no reason to believe that a public natural gas system could not do the same. Moreover, the Conservative Party was still proud of its public power legacy, however much they may have criticized Ontario Hydro from time to time. Premier Leslie Frost, who succeeded George Drew in 1949 and went on to win three successive majority governments, had the unalloyed opportunity to join Premier Whitney and Adam Beck in the pantheon of public heroes. He was reportedly even considering a publicly owned natural gas system in the early-1950s.

In the end, tragically, Frost turned the entire industry over to private interests. The rationale offered for this decision by Frost's Provincial Treasurer, Dana Porter, was both bizarre and preposterous. He said it would be unfair for government to compete with privately owned coal and oil distribution companies, who would lose market share to lower-priced natural gas. Never mind that consumers would benefit from the lower prices. Never mind that moving to natural gas would keep more of Ontario's money in the Canadian economy. Never mind that displacing privately owned coal generators had been one of the major objectives of the public power movement. Why the pangs of conscience about competing with the private sector now? Thankfully, Porter had not been around to advise Premier Whitney in 1905.

Led by the inexhaustible Donald C. MacDonald, the Ontario CCF (Cooperative Commonwealth Federation, predecessor of the New

Donald C. MacDonald led the Ontario NDP, and its predecessor, the CCF, from 1953 to 1970. He retired from the Legislature in 1982. In the mid-1950s, "The Happy Warrior" fought for a public natural gas system similar to the public Hydro system. Instead, the Conservative government of Leslie Frost surrendered the natural gas industry to private interests, a decision that haunts Ontario to this day. *Ontario NDP Caucus Services*

Democratic Party), fought tooth and nail, inside the Legislature and out, for a publicly owned natural gas distribution system, but to no avail. By Ontario Cabinet decision, private natural gas monopolies were established in Ontario a half century after private power monopolies had been dismantled by popular vote. This time, however, there was no vote. The opinion of the people of Ontario was not sought. Perhaps this was because high-ranking members of the Conservative party, and even a handful of Cabinet ministers, had personal interests in the companies that stood to profit enormously if the distribution sector was made a private domain.

Meanwhile, in Ottawa, the Liberal government of Prime Minister Louis St. Laurent had been tag-teaming with Frost to kill the idea of public ownership of a cross-country natural gas transmission pipeline. In 1951, St. Laurent's Minister of Trade and Commerce, Clarence Decatur (C.D.) Howe, travelled to Texas to meet with Clinton Williams Murchison, who had made a fortune in West Texas oil leases and controlled an empire of more than 100 companies—a fossil fuel version of Samuel Insull. Murchison saw even greater wealth in the Alberta gas fields. He set up TransCanada PipeLines with some minority Canadian partners and told Howe he would raise all the money for the east-west pipeline himself. Howe agreed with the plan and TCPL was given five years to raise the $375 million needed to begin construction on the main line.

By the spring of 1956, TCPL had to admit that it was nowhere near raising the money. It said it could finance the 925-kilometre Prairie section and the southern Ontario link because it had raised $40 million from the sale of common shares, $80 million from debentures and $144 million in first mortgage bonds. Missing was the more than $118 million needed for the most difficult and unprofitable section, the 1,100-kilometre leg across the muskeg and forests north of Lake Superior. TCPL executives went to Ottawa, stetsons in hand, to ask for a government guarantee of its debentures. Howe agreed to the public taking on the risk, provided TCPL gave the government enough shares to have a voice in the company. The Canadian front men for TCPL agreed but the American majority interests said absolutely not, contemptuous of the very idea of a "foreign" government having a say in their company.

At that point, the Liberals could have simply let TCPL's concession expire. They could have easily financed the entire pipeline at Government of Canada rates. Instead, Mitchell Sharp, C.D. Howe's assistant, proposed creating a Crown corporation to build only the north of Superior stretch. Howe loved the idea and asked Parliament to approve the Northern Ontario Pipe Line Corporation and give it the authority to finance and build the line. He also agreed with TCPL that they could later buy that section from the government (at less than it cost to build!). Finding the money to buy out the government would be no problem. TCPL would simply use the steady stream of revenues it would collect by selling eastern Canadians western gas at a healthy profit. In Ontario, Premier Frost pledged to join the federal Liberals in financing this subsidy to U.S. gas and financial interests. Ontario taxpayers would have the privilege of paying taxes to two levels of government in order to finance the most expensive option for the delivery of Western gas to Ontario consumers.

It gets worse. In April 1956, TCPL said: "Umm...that bit about the $80 million in debentures we told you we had? Not true. We don't have it. Sorry. What do we do now? Perhaps we should postpone the project." Howe would not hear of it. A pipeline there shall be, and a privately owned one at that, even if it had to be financed by the public. Howe, famous for his "What's a million?" parliamentary riposte, offered the Texas company a government grant to cover the $80 million, but with a condition. The $80 million, plus five per cent, had to be repaid by April, 1957. If not, the government would take over the line for 10 per cent less than it cost. Howe publicly presented this as astute negotiating on behalf of the Canadian taxpayers, but it was no such thing. Private American money would be more than happy to take out the Canadian government stake in return for a guarantee that the pipeline would be built. Why? Because their risk became negligible. In any large construction project, your main risk is non-completion

In 1935, M.J. Coldwell was fired after 20 years as principal of a Saskatchewan school for running for Parliament under the Cooperative Commonwealth Federation (CCF) banner. He won the election and remained in the House of Commons for 23 years, most of them as the Federal CCF Leader. Although Coldwell was a key figure in the parliamentary revolt against the Liberals' gift of Canada's natural gas industry to American interests, he lost his seat in the Diefenbaker sweep of 1958.

because financing falls through for some unforeseen reason, such as a partner's sudden inability to come up with their share. The Canadian government, however, was clearly good for the money, never having defaulted on a loan. It was a sweetheart deal that had to have been made at the very top levels of the government and financial interests involved. Either some people (or the Liberal Party) got paid off in this deal or there were some incredibly stupid people in that government.

Since they had always got what they asked for from the Liberals, TCPL then said thanks, but they simply could not meet the repayment deadline unless the money was approved by June 7, 1956, which would give them ten months of free access to $80 million. At that point, the reliably compliant Howe had just two weeks to get parliamentary approval for the total of $198 million that would go towards the pipeline. What followed was the most fractious episode in the history of the Canadian House of Commons: The Great Pipeline Debate.

Both the federal CCF, led by M.J. Coldwell, and the John Diefenbaker-led Conservatives used every procedural rule available to stall the legislation. In retaliation, the majority Liberals invoked closure, putting a time limit on debate. The opposition appealed to the Speaker, Louis-René Beaudoin, a Liberal. On Thursday, May 31, 1956, Beaudoin ruled in favour of the opposition, renewing their fight to delay the bill. The next morning, however, Beaudoin announced that he had made a mistake and reversed his decision. Rumours were strong that Beaudoin had been visited by Liberal House Leader Walter Harris and called by Immigration Minister Jack Pickersgill, but both men denied influencing the Speaker's decision. Coldwell, beside himself with rage, approached the Speaker and shook his fist at him, calling him a coward and a traitor. A censure motion against the Speaker was demanded, an unprecedented event. Beaudoin offered to resign, but St. Laurent persuaded him to remain.

For the next six days, pandemonium reigned. When opposition members rose to speak, Liberal backbenchers started singing to drown them out. Beaudoin, who left Parliament in 1958 and eventually showed up as a bartender in Arizona, could not keep order. On June 6, 1956, the pipeline bill was passed. Senate approval was granted the next day, meeting the American company's deadline.

The pipeline sellout was the main reason for the Liberals' defeat in 1957 by Diefenbaker's Conservatives. C.D. Howe was unseated in his Thunder Bay riding by a CCF candidate. But even though a 1958 federal Royal Commission on Energy appointed by Diefenbaker quantified the magnitude of this great economic betrayal, he ultimately did nothing to reverse it. The only vestige of Canadian autonomy he allowed was federal regulation of the wellhead price of natural gas. That regulation was scrapped by the Mulroney Conservative government in 1985 with the signing of what has become known as the Halloween Accord.

While it's true that the price of Western natural gas initially went down after Halloween, 1985, this wasn't done for the benefit of Canadians, but rather American gas marketers. Gas exports to the U.S. have quadrupled since then. By 1999, Western Canada gas had 14 percent of the American marketplace. This was also the year natural gas prices began soaring, eventually reaching four times the pre-deregulation price. Prices settled down somewhat after 2001 but still average more than double what they were at the end of 1998. We used to have relatively cheap gas at stable prices. Now we have roller-coaster prices, sometimes lower than the previously regulated price but often higher and sometimes much, much higher. This is a natural consequence of deregulated energy commodities, as electricity customers in Ontario quickly discovered after the deregulation of our power market on May 1, 2002. In the place of stable electricity rates among the lowest in North America, Ontario experienced wildly fluctuating rates that averaged out to well over the pre-deregulation price. Just like natural gas.

We are now left to deal with the lethal legacy of both the Conservative and Liberal decisions to give away our natural gas industry to private, mostly foreign, interests and deregulate what these foreign companies charge us for our own resources. The high and increasingly volatile gas prices and the billions of dollars drained from Ontario's economy over the years are only part of that legacy. The belching smokestacks of our coal-fired generating stations, the growing pile of radioactive wastes that must be buried and secured for thousands of years—assuming we can find a place to do this—and the leftover Ontario nuclear debt of tens of billions of dollars, can also be traced back to these unbelievably short-sighted decisions.

How is it that a privatized natural gas industry led to the rampant construction of large coal-fired and nuclear plants? The answer lies in the intense competition that ensued between electricity and natural gas when gas first reached southern Ontario markets. In a head-to-head competition, gas has some inherent advantages. Of all the energy consumed in an average Canadian residence, fully 85 percent is used for space and water heating. In commercial buildings, it's 62 percent. Even though it was more expensive than it needed to be, natural gas coming into Ontario could still perform these heating functions more cost effectively and more efficiently than electricity, even electricity at cost. Ontario's power industry panicked. Natural gas was going to slash its existing market share and stifle its attempts to edge out coal and oil for heating. Something had to be done to give electricity some advantages. Ontario Hydro, egged on by the municipal electric utilities and manufacturers of electrical equipment and appliances, fought the natural gas invasion with a 20-year program of artificially pumping up the demand for electricity.

It need not have happened. An integrated public power and natural gas system would have allowed government to rationally divide the energy marketplace between electricity and gas and give to each its most efficient role: electricity for lighting and motors, natural gas for heating and cooking. (The use of natural gas to generate electricity was only in its experimental stages back then. The first gas turbines used to generate electricity for utilities went into service in the U.S. in 1961.) But rationality and truth were early casualties in the anti-gas war waged by Ontario Hydro under the soon ubiquitous battle cry: "Live Better Electrically."

It was a one-sided war. The economic and operational advantages of natural gas were no match for the Ontario Hydro propaganda juggernaut. There was, first of all, the fact that virtually every residence in the province was already a Hydro customer. It cost next to nothing to insert promotional and "educational" material into regular hydro bills. These

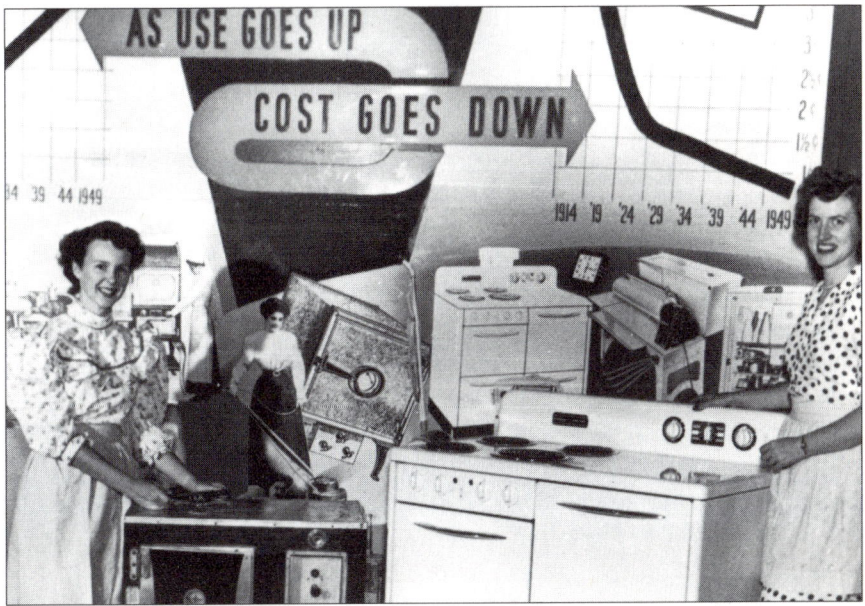

Two-faced Hydro. In 1947, when post-war power demand was soaring, Ontario Hydro implored customers to use less electricity. Just three years later, at the Canadian National Exhibition, Hydro was imploring customers to use more. Janet Kingstone (left) and Edith Kent cheerfully explained the "more costs less" theory, one of Ontario's best examples of delusional economics, right up there with deregulation and privatization. *Hydro One Archives*

bill stuffers were typically tied to wave after wave of print and television advertising campaigns for the all-electric lifestyle. A reincarnation of Adam Beck's Hydro Circus hit the road: an elaborately converted tractor-trailer chock full of gleaming electrical devices and appliances, from refrigerators, freezers, clothes dryers and water heaters to perennial wedding gift favourites such as carving knives, countertop skillets and toaster ovens. Electric appliances were donated to school home economics departments, some of which were visited by "Hydro's Homemakers." At the Canadian National Exhibition, nearly 10,000 people a day visited the all-electric Gold Medallion "dream home" that was staffed by vivacious, entertaining demonstrators. Tens of thousands of Gold Medallion homes were built between 1960-75; they used up to five times the electricity of conventional homes. Exterior floodlights were standard equipment, ostensibly for "safety" purposes. The once-proud owners of these houses found them difficult to sell after the nuclear-related hydro rate hikes of the 1970s turned them into electrical albatrosses.

Hydro had other early advantages in its war on natural gas. Many areas of the province, including some cities, had little or no gas distribution capacity. It was difficult to get private sector financing to build such systems when it looked like they would be unprofitable for a long time, perhaps forever. Ontario Hydro and the locally-elected hydro commissions also had a vast reservoir of public trust to draw upon. For over half a century "The Hydro" had been a central feature of life in Ontario and even with its political problems from time to time, was still seen as the people's power company. Most important, though, Ontario Hydro had the ability to wage price wars, and stupidly did so. In the early 1960s, metered rates were reduced by 10 percent. Even more astonishing, electric space and water heaters were taken off meter and charged flat rates. If you turned your home into a 24-hour-a-day sauna, you paid the same flat rate as someone who used energy wisely. Industrial rate structures also discouraged energy conservation or efficiency. The more electricity you used, the less you paid per kilowatt hour. This may have made sense in Adam Beck's day, when increased demand actually did result in lower rates because it spread out Hydro's initial high capital costs over a wider base of consumption. But it was positively perverse in the late 1950s and early 1960s, when Hydro once again faced the prospect of energy shortages, particularly during the long winter heating seasons. Except for the northern part of the province, far from major population centres, Ontario was running out of larger scale hydroelectric potential.

The seemingly endless Live Better Electrically campaign did a smashing job of repressing the adoption of natural gas for more than 20 years. But it was a pyrrhic victory that haunts us to this day. Demand for power did soar, taking the province to the brink of blackouts during many winters in the 1960s. Coal-burning plants were rushed into service. The nuclear program was accelerated at great cost. Even though the untested CANDU heavy water reactor design was still in development, Ontario Hydro committed in the early 1960s to building large nuclear stations on Lake Huron and Lake Ontario. Assurances that power from these stations would be "too cheap to meter" have proven, as we now know, ludicrously wrong.

Hydro has long been painted as a demonic, empire-building force unto itself, an out of control behemoth that is protected from the consequences of its excesses by its monopoly. To be sure, the Hydro-conceived, Hydro-executed, self-reinforcing death-spiral of promoting demand to make the most use of capacity, and then having to build more capacity, at an ever-increasing cost, to meet the rising demand, gives a lot of credence to this view. I have voiced it many times.

But I was also part of a government that *did* rein in the beast. Premier Rae took the position from day one—October 1, 1990—that we represented Hydro's shareholders, the people of Ontario. We took charge of the situation by appointing highly qualified Chairs and board members that shared our vision of the future. We ordered a moratorium on nuclear construction. We shut down Ontario Hydro's Construction Division, which existed for only one reason: to build more nuclear generating stations. We approved the largest downsizing of a publicly owned enterprise in Canadian history. We enhanced the authority of the Ontario Energy Board. We put the brakes on Hydro's growing debt and reversed it. We froze rates because of the deep recession of the early 1990s, mainly to prevent smaller, more economically vulnerable energy-intensive industries from going under. We sought to redress legitimate First Nations grievances against Ontario Hydro. We ordered much more attention be paid to energy efficiency and the environment. Not a single kilowatt of coal or nuclear capacity was approved throughout the NDP years, but $110 million was committed in 1994 for renewable energy technologies.

At the end of our mandate, in 1995, Hydro's net income was 207 percent higher than in 1991, the first full year of the Rae government. In-service capacity was 6.5 percent less, but still more than enough, at 29,244 megawatts, to meet 2002's record peak of about 25,600 megawatts. The long-term debt had gone down nearly three-quarters of a billion dollars—faster than had been projected—and the debt ratio had declined for the second year in a row. Ontario Hydro had begun to pull itself out of its downward spiral and was starting to devote itself to promoting energy efficiency and sustainable development. Things were looking up for the people's power company. In the 1995 Ontario Hydro Annual Report, the President and the just appointed Conservative Chairman jointly declared: "Revenues remain strong; the debt is coming down faster than projected, costs are being reduced and services delivered more efficiently. The major restructuring and management reorganization that began [in 1992] continues, as do the benefits. We are very optimistic about Hydro's future…"

While it's true that the Conservatives took over in the latter half of 1995, they can't take credit for the excellent results of that year, since operational plans and budgets for 1995 were all in place by the fall of 1994. It is an easily verifiable fact that the NDP was turning Hydro around for the first time in a generation, maybe longer. We handed over to the Harris government a thriving public enterprise. In the same message from the Chairman and President in the 1995 report quoted above, they said that Ontario Hydro was "the largest electrical utility on the continent—with very low marginal generation costs and a significant surplus to sell." Despite this clear endorsement of our public power system from a Chairman newly appointed by the Conservative government, the Conservatives to this day claim that Ontario Hydro was a dysfunctional wreck that had to be broken up and subjected to the discipline of the marketplace. In light of the deregulation disaster of 2002, however, it now appears that the Conservatives are the ones who were humbled by the marketplace, and will soon be disciplined by the voters.

I've jumped ahead in the Hydro story from the 1960s to the mid-1990s to underline the fact that lack of political will to govern Hydro was the real reason for the coal and nuclear problems we and our children must now face. The history of Ontario Hydro would have been very different under an NDP government. Even before Western natural gas reached Ontario, our party was calling for a publicly owned natural gas system and an energy policy that integrated hydroelectricity and natural gas. And when we finally won government in 1990, we put a stop to the Conservative/Liberal nuclear madness and began to successfully implement our longstanding commitment to energy efficiency and renewable energy. It didn't take the discipline of the marketplace to bring Canada's biggest publicly owned enterprise under control. It took a government with a clear vision of an economically and environmentally sustainable future and the political will to implement it. We would do so again.

10
Nuclear Ontario

By the mid-1970s, it was clear that the frenzied push for nuclear power was an economic and environmental disaster in the making. Conservative and Liberal governments ignored the warnings.

For at least two reasons, it is easy to understand why nuclear energy was initially such an attractive idea. In theory at least, it offers the most efficient transformation of matter into energy. In keeping with Einstein's mathematical equation $E = mc^2$, one pound of anything, when it is completely converted into energy, will produce the equivalent of more than 11 billion kilowatt-hours of energy. That's the annual use of about one million Ontario households. Even if we were only able to achieve a small fraction of this Einsteinian ideal, we would have limitless amounts of literally dirt-cheap electricity. Secondly, the idea of harnessing nuclear energy for electricity might perhaps help diminish the apocalyptic spectre of "the bomb" that haunted the post-Hiroshima world for decades. We needed to believe that this terrifying force we had unleashed could be converted to some useful, redemptive purpose.

For Canadians, there was a third reason: national pride. In the early years of World War II, the Canadian nuclear research program was as advanced as that of the Americans. At that time, the leading atomic scientists in the world were European. A number of them escaped to England before Hitler's armies, carrying with them the world's entire supply of heavy water: 200 kilograms, a few pailfuls. Cambridge University, Britain's scientific research centre, did not have enough facilities to support nuclear fission research and, in any event, England was a dangerous place to be. A select group of the scientists relocated to McGill University in Montreal.

The theoretical framework for understanding how nuclear fission works was first sketched out by Otto Hahn in Berlin and Lise Meitner, a German Jew who had fled the Nazi regime to work with Neils Bohr in Norway. (Probably because she was a woman, and possibly because she was Jewish, Meitner did not at first get the credit she deserved for her seminal role in nuclear discoveries. History has since corrected this.) At around the same time, a group of French scientists, the Paris Group, was working on controlling fission and mused about the possibility of using it to generate electricity. There was considerable atomic research also going on in the United States and Great Britain. By early 1939, these scientists realized that fission could also be used to make bombs. Knowing that Hitler would not hesitate to use atomic weapons, the Allies raced to produce their own deadly devices. In one of history's most horrifying ironies, the atomic bomb was eventually used only against Germany's ally, Japan, which had little chance of developing such weapons on its own.

Although there was considerable cooperation between Canadian and American nuclear research efforts during the war, each went its separate way afterward. The U.S. focused mainly on making bigger bombs and nuclear-powered submarines. The commercialization of nuclear energy, while supported, was left mainly to the private sector. The technology diverged as well. Canada stuck with heavy water systems, the Americans took the slightly different, light water route.

The Canadians enjoyed early, encouraging results. In 1947, the National Research X-perimental (NRX) reactor at Chalk River, northwest of Ottawa, became one of the first in the world to generate electricity. The NRX also began producing medical isotopes for research and cancer treatment. Radiation therapy is a Canadian innovation, first used at London's Victoria Hospital. In 1949, Ontario Hydro began thinking that it might steal a march on the Americans and be the first to commercialize nuclear energy production. And why not? Canada had the expertise and the technology, which was continuously advancing. Ontario needed the power; demand was rising fast and the province was running out of exploitable large scale hydraulic sites. Moreover, we had great uranium deposits in the north and a well-run refinery at Port Hope in the south. Coal, on the other hand, had to be imported. When all these factors were added up, going nuclear looked very good on paper. Just like deregulation did, fifty years later.

The first sign of trouble came on December 12, 1952, when a uranium fuel meltdown at Chalk River blew the four-ton lid off the NRX reactor and released significant radioactivity into the station and the surrounding air—the world's first major nuclear accident. Thankfully, the building was evacuated in time and no one was injured. So great was the damage, however, the U.S. was eager to take a look. Future U.S. President and Nobel Peace Prize winner Jimmy Carter, who had been trained as a nuclear engineer, was one of the American servicemen rushed north to help the Canadians clean up.

The commitment to nuclear power survived the explosion. Work on the NRU, another reactor at Chalk River, continued as planned and even larger projects were enthusiastically pursued. Ontario Hydro and the federal Crown corporation, Atomic Energy of Canada Ltd. (AECL), began working on a 25 megawatt demonstration project at Rolphton, upstream from Chalk River and a 200 MW reactor at Douglas Point, on Lake Huron. Douglas Point was to be the first CANDU reactor. It was also Ontario's first nuclear money pit. In 1957, after four years and tens of millions of public dollars, the original CANDU design was tossed out as unworkable and a brand new one proposed. Douglas Point lurched along for eleven more years before it began producing commercially reliable power in 1968. Meanwhile, in 1958, another accident had occurred at Chalk River, this one at the newer NRU reactor. A uranium rod fell on the floor and caught fire, causing significant, but contained contamination. It was much less serious than the NRX accident, so the Americans stayed home this time.

Hydro's dreams of being the first to sell nuclear power were dashed when the world's first commercial nuclear station, a 73 megawatt reactor at Shippingport, Pennsylvania, went on-line in 1957. But if we could not be the first, maybe we could be the biggest. In the early 1960s, Ontario Hydro announced that Canada's first large-scale nuclear reactors would be built on Lake Ontario, at Pickering, just east of Toronto. The original plan for two 540 MW reactors was quickly scaled up to four. This was the Pickering A station, which began generating power in 1971. So confident were Hydro and AECL engineers in the design of Pickering that they began working on four more reactors at the site (Pickering B) and eight even larger reactors (Bruce A & B) on Lake Huron, along with a heavy water plant. By the mid-1970s, more than 10,000 MW of nuclear power capacity was either in service or in the final planning stages. Moreover, Hydro had already begun planning the giant 4 X 900 MW reactor Darlington station, just east of Pickering. It would be bigger than any nuclear facility station in the U.S. at that time.

As big as it was, however, Darlington would barely make a dent in what Ontario Hydro predicted would be the province's future energy needs. In early 1974, Hydro unveiled its long-term plan for meeting

Ontario's seemingly inexhaustible appetite for power—an appetite Hydro itself had been furiously stoking for nearly 20 years. The gargantuan plan called for more than 150 generating units to rim the Great Lakes. Had it gone ahead full steam, no pun intended, it would have been literally breathtaking, as almost half the units were to be coal-fired. By 2000, Hydro said, Ontario would need "approximately" 90,000 megawatts. Actual Ontario peak demand in 2000 was under 25,000 megawatts.

The soaring ambition of Ontario Hydro (and AECL) engineers knew no bounds and certainly knew no restraints. Setting boundaries and restraints, determining Ontario's energy policy, that was the government's job. The Conservatives, however, who had been in power for three straight decades and would have presumably figured out in that time where the levers of authority were, seemed incapable of governing Ontario Hydro. Or maybe they were just unwilling to do so. After all, a lot of the right people with the right connections benefited enormously from this unbridled construction activity. Take Babcock and Wilcox, for example, the U.S.-based supplier of generating station boilers for both coal and nuclear units. According to one account, B&W had botched the manufacture of boilers for Pickering A and B, costing Hydro hundreds of millions of dollars and lengthy construction delays. When asked to take some financial responsibility for the mess, B&W threatened to close its Ontario operations in Cambridge and put the 1,600 workers on the street. Hydro capitulated and picked up most of the tab, which by 1982 had grown to $850 million. The decision to let B&W off the hook had to have been approved by the Davis government. Indeed, the Conservatives had already actively interfered in the awarding of the boiler contracts for the Bruce A station to B&W, against the advice of both AECL and Ontario Hydro staff.

Then there's the astonishing story of the sordid politics behind the multi-billion dollar uranium contracts. In the early 1970s, Hydro began looking for a long-term supply of cheap uranium to fuel the dozens of reactors it was planning to build. Hydro management recommended simply taking over Denison Mines in Elliot Lake and paying the owner a fair price. Premier Davis and his Treasurer, Darcy McKeough, didn't like the idea of public ownership of mines and told Hydro to come to an agreement on long-term contracts with both Denison Mines and nearby Rio Algom Mines. After long negotiations, contracts that would ultimately cost Hydro somewhere between $6 billion and $7 billion were arrived at in 1977, but not signed right away. Davis was in a minority government and wanted to spread political responsibility for the huge decision, the largest single contract ever awarded by Ontario

Pickering Nuclear Generating Station, on Lake Ontario, just east of Toronto. Energy that was supposed to be "too cheap to meter" has proven very costly. As of March 2003, four of Pickering's eight reactors had been shut down for six years by the Conservative government. The cost of the reactors' rehabilitation, like their original construction cost, has soared well above projections. *Hydro One Archives*

Hydro. He referred the matter to the all-party Select Committee on Ontario Hydro, hoping to get the Liberals on side. He knew the NDP would never agree. It is important to this story that Stephen Roman, the owner of Denison Mines, was a personal friend of Davis.

The NDP adamantly opposed the contracts, calling for a government takeover of the mines. After all, Hydro had a past history of buying out private power companies, all with the approval of previous Conservative governments. This was simply an extension of that practice. It would also be a better hedge against future uncertainty. The NDP was opposed to building more nuclear plants, but was in no position yet to stop them. Let's at least take control of the fuel resources, the NDP argued. If sanity later prevails and we *do* halt the nuclear march, Hydro won't be stuck with long-term, high-priced contracts. And besides, how confident are we of the world price for uranium for the next three decades?

With their historic commitment to private power, the Liberals had no problem with the contracts in principle, but nitpicked at the price to try to embarrass the Conservatives. Taking both sides of a controversial issue remains the Liberals' political strategy to this day. As we shall see later, with the final Darlington decision, this approach proved very costly for Ontario.

In February 1978, after weeks of clamorous hearings, the Liberal and NDP members of the committee outvoted the Conservatives and rejected, for their own different reasons, the proposed Denison and Rio Algom contracts. Davis had lost his gamble. Instead of conceding defeat, Davis overruled the committee and ordered the contracts signed, guaranteeing Stephen Roman over a billion dollars in profits. It proved to be a horrible blunder, even worse than the NDP had feared. At the time, world uranium prices were just over US $43 per lb. The 30-year Hydro contracts were for about $60 per lb. Perhaps if our nuclear reactors had come within shouting distance of Einstein's energy conversion ideal, this heavy premium would have been no big deal; we would have only needed around 15 pounds of the stuff every year. But the theory remained theory; we would actually need much more uranium than that. The contracts were for millions of pounds a year.

Within months after the contracts were signed, the world price of uranium started dropping. By 1985, before a single ounce of uranium had been delivered, prices had dropped to $15, one-quarter the contract price. By October 1990, when the NDP took over, world uranium prices had dipped below $10, where they have stayed since. A few months later, in April 1991, Premier Bob Rae ordered Ontario Hydro to finally cancel the contract. Denison claimed over $360 million in damages, but an arbitration panel awarded it just over $30 million. Cancellation of the Conservative contracts saved Ontario Hydro billions of dollars, something the Conservatives of today neglect to mention when they lob vague accusations of fiscal imprudence during the NDP government years across the floor of the Legislature.

The Ontario New Democratic Party was not always officially opposed to nuclear power development, though many in the party were. In 1970, a resolution passed at its convention took a "let's see what this is all about" stance. The reasons were straightforward: consciousness of the environmental impacts of pollution from coal plants was growing. Maybe Hydro knew what it was doing. There hadn't been any nuclear mishaps in 12 years. And besides, there were quite a few jobs, good unionized jobs, in nuclear plant construction and uranium mining.

As the decade wore on, however, the awful truth emerged. The huge cost overruns and long delays that had plagued the Rolphton and Douglas Point plants were not just growing pains. Rolphton took four years longer and twice as many dollars to build as projected. Douglas Point eventually came in at nearly 50 percent over budget and never worked properly. It limped along at less than half its alleged capacity for nearly 20 years before being permanently shut down in 1987. These experiences turned out to be typical. Pickering A was 50 percent over budget, not counting the boiler repairs due to Babcock and Wilcox's screw-up. The Bruce Heavy Water Plant cost double what had been projected. Cost overruns and cancellations added over $700 million to the final tab. The same sad saga would eventually be told of the Bruce A, Pickering B and Bruce B plants. At the same time, beginning in the early 1970s, demand growth began levelling out, renewable energy technologies were becoming more sophisticated and affordable, energy efficiency was a growth industry. By the time plans for the Darlington station were announced in the mid-1970s, the NDP had turned solidly against building any more nuclear plants.

To fully understand how Ontario's electricity system came to be in the mess that it is in today, two of Ontario's major energy stories must be told. The first is the natural gas betrayal. The second is the Darlington mega-fiasco. The two are closely related. Darlington is the offspring of the Frost Conservatives' decision to give the natural gas industry to private interests and the subsequent energy civil war in which Ontario Hydro went to insane lengths to boost electricity demand and prevent natural gas from gaining market share.

Ontario Hydro was always good about announcing big projects when they were still on the drawing boards, often before. This was politically motivated, naturally: it's always better if the public is given lots of notice about large public expenditures, it gives the appearance of democracy. By 1974, we knew for sure that Darlington was coming our way. The NDP wanted to head it off at the pass, calling for a moratorium on any more nuclear construction. We were convinced that it was unnecessary and we worried about the immense environmental impacts of a second nuclear station on Lake Ontario, not to mention the still-unresolved nuclear waste issue. In the Legislature, we demanded that the project be subject to the newly passed Environmental Assessment Act. The Conservative government of Bill Davis, however, said no. The government's rationalization was absurd: "Planning for Darlington is already several years down the road. You can't hold up a project that far along because of a law that didn't exist when the project began. An environmental assessment could take as long as two years. We can't afford the delays. Maybe next time." That was in 1974. Fifteen years later, Darlington was still under construction.

Two months before the 1975 election, Premier Davis swept all issues about Ontario Hydro off the political agenda by appointing Dr. Arthur Porter to head up a five-person Royal Commission with a broad mandate to study and make recommendations on every aspect of the province's electricity system. The end result, as we shall see, turned out to be a pretty good road map to the best energy future for Ontario.

Davis was undoubtedly concerned about Ontario's energy future, but he was also worried about his immediate political future. Hydro rates had begun rising sharply and Davis had been given forewarning of a 1976 rate increase of 30 percent, thanks largely to Pickering. Moreover, the economy was in rough shape, with runaway inflation hammering the budget. Labour unrest over the province's antiquated health and safety laws dogged the government, along with a raft of social and environmental issues: education, health care, high rents, acid rain, you name it. The mid-1970s were a difficult time for most governments in the Western world and the eloquent Ontario NDP leader, Stephen Lewis, made it especially difficult for the Davis Conservatives. The Porter Commission was an attempt by Davis to skate right past energy issues and perhaps even be seen to be doing something about them while, in fact, doing nothing. The beleaguered premier could then, he hoped, focus on defusing the other issues and win his second majority government. He could then wind up the Porter Commission quickly, picking and choosing which of its recommendations to implement.

It didn't work. On September 18, 1975, Lewis led the NDP to Official Opposition status, three seats ahead of the Liberals. The Liberals had failed to capitalize on dissatisfaction with the Conservatives, probably because they were indistinguishable from them. One of the best outcomes of the minority government, besides a greatly improved occupational health and safety law, which has indisputably saved many workers' lives, was the establishment of a Select Committee on Ontario Hydro. The committee was chaired by former NDP Leader Donald C. MacDonald, who had lost the battle for a publicly owned natural gas system 20 years earlier. Time had not eroded his fervour for public power, but experience had taught him that Ontario Hydro needed to be held on a tight leash and quizzed often, and in depth, on its planning assumptions and its operations. The veteran MacDonald was backed up by an NDP rookie member, Carleton East MPP Evelyn Gigantes, who quickly became Darlington's most implacable political foe. A visibly pregnant Gigantes once demonstrated in front of the station while

New Democrat MPP Evelyn Gigantes was an eloquent opponent of Darlington nuclear station from her first election in 1975. This demonstration was in 1980; it would be almost 13 years before the station was completed. *Alex MacDonald*

it was under construction. Her hand-lettered sign read: "Nukes Produce Killer Watts."

The Select Committee concluded in 1976 that no nuclear plants were needed beyond Pickering and Bruce. Hydro's demand growth projections were preposterous. Despite its rigorous analytical underpinnings, however, the committee's recommendation was ignored. Davis called an election for June 1977, but failed once again to get a majority. This time the Liberals squeaked past the NDP by one seat to regain Official Opposition status. One month later, Davis gave the order to start Darlington.

The Select Committee continued to investigate Hydro issues until 1981, when Davis finally won a majority and disbanded the pesky, inquisitive group. But the report of the five-year-long Porter Commission, appointed by Davis himself, could not be so easily dismissed. Its very first recommendation was that "demand management" should guide future power system planning rather than the traditional focus on supply expansion. Stop building and start conserving, in other words. Indeed, the report was replete with concrete expressions of a completely new approach to energy development and use. While it mistakenly thought that at most one more nuclear station should be built after Darlington, the report warned at length about nuclear energy's unexamined hazards and hidden costs. Among them: poisonous uranium mine tailings, occupational cancers, the lack of nuclear waste disposal solutions and the limitations of human beings in dealing with such phenomenally complex technology. Porter called for heavy support for biomass, solar and other low impact generation, as well as an intense campaign to greatly improve industrial energy efficiency. The report also recommended much stricter efficiency standards for all major energy-consuming appliances. For the most part, the report validated NDP energy policy, including our concerns for significantly reducing fossil emissions. Michael Cassidy, who succeeded Stephen Lewis as NDP leader in 1978, made implementation of Porter's 1980 reccommendations a major issue in the 1981 election. He also vowed to cancel the scandalous uranium contracts, a promise finally fulfilled by Premier Rae ten years later.

Did I say that the Porter Report couldn't be easily dismissed? I must have been thinking about how a responsible government would have acted after being handed the results of the most expansive and in-depth study of the province's electricity system since the days of Adam Beck. Davis, however, studiously ignored it. Government officials would stare blankly at its mention. "Porter? Porter? Ah, yes, Porter. Very forward-looking chap, eh? A bit *too* forward-looking though, wouldn't you say?" It wasn't until a quarter-century later that a desperate Conservative government, looking for a way to divert attention from its Hydro privatization and deregulation fiasco, finally decided that perhaps renewable energy and energy efficiency were good things that deserved public support. It was time, they said, to be more forward-looking.

The May 1985 election was a watershed in Ontario political history. After 14 years as Premier, Bill Davis announced his retirement in early 1985. He could read the polls better than anyone and had no interest in another minority government, or worse, an election defeat. Amiable, plaid-jacketed Frank Miller won the leadership and then went on to prove Davis right. His party won the most seats, 52, but was 11 shy of a majority. The Liberals, led by London's David Peterson, trailed the Conservatives by four seats but won 38 percent of the vote, a point more than the Conservatives. The NDP, under first-time Leader Bob Rae, captured 25 seats, losing a number of close races to the Liberals.

Both Peterson and Rae were eager to put an end to 42 years of Conservative rule. The fact that the Liberals had won the largest percentage of votes made the thought of a compact between the two parties politically defensible. And that is what happened, with intense negotiating between the parties producing a minority government pact (the "Accord") that put the Liberals in power for at least two years. The agreement was unprecedented. It set out an agenda for reform that opened up government processes to public scrutiny and introduced many legislative, social and economic reforms. "Extra billing" by doctors would be banned and medically necessary travel for Northern Ontario residents would be covered under OHIP. An ambitious non-profit housing program would be launched, along with reforms to the patchwork and largely ineffective tenant protection laws. Workers would be given the right to know about toxic substances and other workplace hazards they faced. Programs to create employment and training opportunities would be introduced. Freedom of Information and Protection of Privacy would be enshrined in law. In the middle of a long list of political reforms, including televised Legislature proceedings, campaign spending limits and more power to the Provincial Auditor, there was this non-negotiable NDP requirement: "A Standing Committee on Energy to oversee Ontario Hydro and other energy matters."

Peterson was sworn in on June 26, 1985, the first Liberal Premier since 1943. The Legislature met on July 2. No one wanted to take summer holidays when history was in the making. One week later, the Select Committee on Energy was appointed and told to "report within ten months on Ontario Hydro Affairs," the broadest possible mandate. Darlington was the most urgent matter. Stopping it had been a prominent NDP platform issue. With characteristic frankness, the Liberals had expressed their feeling. Perhaps, they said, if all the facts were to come out, finishing Darlington might not be, strictly speaking, necessary right now, given all the circumstances as we understand them and subject to a thorough review, of course. At least you always know where the Liberals stand.

When the committee began its hearings, construction of Darlington's first two units was already well underway, eight years after the station had been approved by Davis. Work was set to begin on units 3 and 4. Let's review the numbers; they're important to understanding the lunacy of this single largest public project in Canadian history.

Darlington was originally budgeted at just under $4 billion. By mid-1985, the price tag had ballooned to almost $11 billion. That's a thousand dollars for every man, woman and child in the province. More than half had already been spent or committed. Cancelling the project would save $4 billion, but would "waste" the money already spent. Would it really be wasting money, though, not to finish a plant that was never needed in the first place? For anyone who has made it through Economics 101, the answer is simple: along with the balancing benefits, incremental costs, not sunk costs, are the best basis for making a decision about resource allocation. And it was this issue of a sunk $7 billion vs. an incremental $4 billion that was the main question facing the committee, although there were other environmental and safety issues as well.

Was the 3,500 MW of extra power needed? At the time, Ontario Hydro already had 27,300 megawatts of electrical capacity, 40 percent more than 1985's peak demand. By 1992, even without Darlington, this capacity would grow to 30,500 MW, since other, smaller plants were under construction. By 1998, *without* Darlington, Ontario Hydro would be able to deliver 33,100 MW. With Darlington's 3,500 MW added in, total capacity would be 36,600 MW. This included the potential for importing a small amount of power, mainly from the U.S. In addition, Quebec and Manitoba both had large surpluses of very low cost hydroelectric power they would have been eager to sell, at significantly less than the cost of new nuclear. In 1985, Hydro's estimate of anticipated peak demand in 2000 had dropped to 30,000 MW, a forecast that did not presume any serious provincial efforts at conservation or energy efficiency during the next 15 years. The 30,000 estimate proved to be high, but not outrageously so. Peak demand in 2000 hovered near 24,000 MW a few times. During the summer of 2002, it reached an all time high of 25,600 MW. We didn't need Darlington, not for a long time.

The committee issued its report in December 1985. It had heard from 150 witnesses. The report recommended Darlington units 1 and 2 be completed, but that construction on units 3 and 4 should be put on hold for six months, while further study was undertaken. The NDP members of the committee, Brian Charlton and Ruth Grier, wrote a compelling three-page minority dissent:

> There is no justifiable reason to continue the construction of Darlington. Both the Report of the Select Committee on Energy, and the evidence presented to the Committee support this position. The electricity will not be needed until the next century and Darlington is not likely to minimize the future cost of electricity.

Darlington Nuclear Generating Station, on Lake Ontario, east of Pickering. Miscalculations by the Davis Conservatives and Peterson Liberals led to $14.3 billion in new debt for a station Ontario did not need. A fraction of that money spent on energy efficiency would have reduced demand by much more than Darlington can produce. *Hydro One Archives*

Instead, building Darlington will effectively curtail the development of alternative sources of electricity and the implementation of electrical efficiency. Although $7 billion have been committed to Darlington, at least $4 billion, and possibly more, can still be saved.

The committee picked up its work in early 1986 and delivered its final report in July, as required by its mandate. Reading that report today makes one wonder: what were they thinking? Let me summarize its key recommendations concerning supply and demand:

• Ontario Hydro must develop a comprehensive conservation strategy and fund it adequately (to reduce demand).
• The Ministry of Energy must establish high energy efficiency standards for appliances and buildings (to further reduce demand).
• Independent, non-Hydro generation options should be

Brian Charlton and Ruth Grier spent much of the last half of the 1980s trying to stop the Peterson government from spending over $4 billion to complete Darlington. They were ignored by the Liberals and the additional costs ballooned to $7 billion. Charlton and Grier later served as Ministers in the Rae government and launched major energy efficiency and environmental protection programs. *Courtesy of Brian Charlton* (Charlton) *and NDP Caucus Services* (Grier).

encouraged—cogeneration, municipal solid waste and small hydro among them (to increase supply).
• And, even though we don't really need its power, Darlington should proceed despite the need for a more balanced system.

Three months earlier, in the middle of the committee's hearings, an explosion at a nuclear station at Chernobyl in Ukraine, blew off the reactor's massive steel and concrete lid. Thirty people died immediately; 135,000 were evacuated. This is why the committee also recommended the appointment of "An independent panel of internationally recognized experts to review, on a priority basis" Ontario Hydro's nuclear safety and emergency plans. To date, 2,500 deaths are attributed to Chernobyl. Many thousands of premature deaths from cancer are expected.

Once again, Brian Charlton and Ruth Grier dissented from the committee's majority report. Once again, they pointed to Ontario's enormous surplus, noting that some American utilities, also facing surpluses, had mothballed nuclear facilities "at or near the point of completion." Once again, they pleaded for more investment in energy efficiency and renewable generation. Once again, they were ignored.

David Peterson called an election for September 10, 1987, shortly after his two-year agreement with the NDP expired. It was the politically obvious thing to do. The economy was booming. Jobs were plentiful. Toronto's "world class" SkyDome was under construction. The legislative reforms of the last two years had proven popular, particularly the ban on extra billing by doctors and the additional protections for tenants. The government seemed more open, more trustworthy.

It was a massive victory. The Liberals swept 95 seats, 73 percent of the Legislature. It was the second largest electoral triumph in Ontario history. The largest was in 1908, when Premier Whitney won 81 percent of the seats, two years after his government had founded Ontario Hydro. In the midst of the Liberal landslide, I won my first election as MPP for Rainy River. New Democrats dropped six seats in the election but we became the Official Opposition. The Conservatives, led by Larry Grossman, crashed badly, ending up a distant third with only 16 seats, 36 fewer than they had won two years earlier.

Shortly after the election, the Peterson Cabinet bowed to furious lobbying by Ontario Hydro and other interests and gave the green light to finishing Darlington. The first reactor unit began operating in 1990. The final price tag in 1993, when all four units were on stream and delivering power, was $14.3 billion. In order to save us from wasting $7 billion, the Liberals spent another $7 billion. Good thinking: just like Peterson's unfulfilled promise to cover any cost overruns on Skydome.

I have spent so much time talking about Darlington because it lies at the very heart of our struggles with electricity policy today, a quarter century after Bill Davis set it in motion. Darlington—such a nice-sounding name—was the single largest public policy mistake in Ontario history, even larger than the privatized gas industry betrayal. Here's why: if the people of Ontario voted for it today, we could establish a publicly owned natural gas company tomorrow—call it a province-wide gas cooperative. Not a bad idea. But we cannot vote Darlington out of existence.

Darlington added over $14 billion to the Ontario Hydro debt, a debt that has been used to discredit Hydro as a public enterprise and provide the Harris/Eves Conservative governments with their most vivid, albeit flawed, argument for deregulation and privatization, both of which have made the debt even worse. As predicted many times by the NDP, Darlington killed any real interest in energy efficiency and greatly slowed the uptake and further development of renewable energy technologies in Ontario. Moreover, it led directly to the power shortages of 2002.

"Wait a minute!" some of you are thinking, "Where would we be without Darlington today? In the summer of 2002, we had to import thousands of megawatt hours of power from high-priced coal plants in the U.S. Without Darlington's power, we would have had blackouts for certain, yes?"

No. The truth is, it was Darlington itself that drove us to the brink of blackouts. For 16 years, the youngest child of Ontario Hydro's nuclear family got most of the attention and resources. Maintenance and quality control at Pickering and Bruce were sorely neglected throughout much of the 1980s and into the 1990s. We eventually lost more than 5,000 megawatts of generation capacity at Pickering A and Bruce A, due to management neglect and confusion, in order to gain 3,500 MW at Darlington. It only cost us $23 billion, plus interest: $15 billion for Darlington and $8 billion for the (so far) unsuccessful rehabilitation of the older stations. Boy, those Conservatives and Liberals sure know how to run things.

If, instead of Darlington, the government and Hydro had decided to aggressively pursue energy efficiency, we would be much better off today. We certainly wouldn't be needing as much as 26,000 MW, which is less than what Ontario generators have been capable of supplying since deregulation began. For decades, Ontario was a net exporter of electricity. Now we rely on imports to keep the lights on. What a mess! And it's not as if Conservatives and Liberals had no idea that there were alternatives to huge nuclear stations. My NDP colleagues showered them with research that was specific to Ontario. One study commissioned by Bob Rae and Brian Charlton, our Energy Critic, recommended investment in energy efficiency measures that represented just a fraction of the possibilities for painlessly reducing Ontario's energy demand. The proposals would have saved more than 1.5 times the capacity of the entire Darlington generating station when it is humming like a top, which is not always the case, to put it politely. So Davis and Peterson had plenty of information, from us and many others, on the economics of energy efficiency. They made quite a deliberate choice to follow the uranium brick road to the land of "theoretically" cheap power. Now look where we are.

Believe it or not, there's more bad news. Because Ontario took the mega-nuclear route, we will continue to be vulnerable to power shortages for years to come. A nuclear reactor shutdown takes a lot of power out of the system, something Ernie Eves discovered to his dismay in the fateful summer of 2002 when a privatized Bruce Power unit "unexpectedly" went down for several months. Someone should have told Eves that nuclear units are *always* "unexpectedly" going down. It is phenomenally complex technology. There are many things that can go wrong and, in fact, many things do go wrong, often necessitating unit shutdown. In a tight supply situation, in a deregulated market, 850 MW of capacity can be the difference between 4.3 cents per kilowatt hour and double that. Supply and demand. Nothing we can do about it. Who could have guessed? If it wasn't for all those air conditioners…

What if those 850 MW had been distributed around the province in 50 or 100 different small-scale run of the river hydroelectric generating sites? What if, instead of it all coming from a "one-human-error-away-from-meltdown" station, it came from natural gas-fired cogeneration stations, wind turbines, solar cells, biomass, small scale hydro and other renewable sources? With such a diverse, dispersed generation mix (a threat, of course, to the big time multinational players who insist on having a deregulated energy sandbox in which to play their oligopolistic games), we would not be vulnerable to large sudden losses of generation capacity. Nor would we be faced with the choice of either building a large amount of high-cost surplus capacity, to hedge against nuclear shutdowns, or taking the chance that we will be able to import enough power when the nukes go down. Unfortunately, we now have the worst of both choices. We have a large surplus of very expensive nuclear capacity, 50 percent of which was out of commission in the summer of 2002. We had to import very high-priced power during heat waves to make up for the lack of nuclear power that was supposed to be too cheap to meter.

On paper, I admit the renewable power looks more expensive. In the real world, nuclear has proven to be far more expensive and much riskier, even if we forget about the nuclear decommissioning and waste storage costs to come. Regrettably, however, we can't do that, just as we can't vote away Darlington. This is why when I hear the Conservative Minister of Energy *du jour* harp about the huge Hydro debt to be borne by our children—a debt incurred exclusively by Conservative and Liberal governments against the oft-recorded advice of the NDP—I become just a little angry.

There is actually some good news to take away from this sorry tale. We now know, without a doubt, that the crippling problems in Ontario's electricity system—the debt and the shortages—are nuclear-related. These problems are not inherent in public ownership and regulation, as the Conservatives and Liberals falsely claim. U.S. investor-owned utilities have had the same bad experience as Ontario. By getting off the nuclear track as soon as practicable, we can restore and then reshape for a new century what was once, and can be again, one of the world's great power systems.

11

The NDP Years: 1990-1995

The Rae government stopped Ontario Hydro's nuclear madness, changed its focus to energy efficiency, stabilized its escalating rates and reversed its growing debt.

The celebrating began only a half hour after the polls closed on September 6, 1990, when the CBC declared an NDP majority government. And what a celebration it was when Bob Rae and his Cabinet were sworn in and invited "absolutely everyone" to Queen's Park for the first-ever public reception in the Legislature to mark a new government's first day in office. Thousands showed up. For many it was their first time inside the grand rose-hued building. All 74 NDP MPPs, nearly quadruple the party's numbers from the last election, were on hand to greet them.

It was a stunning rebuke for David Peterson's Liberals, who had registered over 50 percent support in pre-election polls, support that began plummeting almost immediately after Peterson called the snap vote. It had been only 33 months since he had won the second largest majority in Ontario history. Fuming editorialists, pulled away from their summer cottages, gave popular voice to a suspicious resentment over an unnecessary election, especially one held just as families were enjoying the last weeks of summer and getting ready for back to school. In what was clearly a case of political punishment, the voters stripped away 59 seats from the Liberals, giving all but four to the NDP. The Conservatives once again came in a distant third. Their new leader, Mike Harris, had run a lacklustre campaign.

We soon learned why Peterson had called an early election. An alarming array of negative economic trends had been converging since late 1989 and were now hitting Ontario like a tsunami. The

Mulroney/Reagan Free Trade Agreement (FTA) of 1988 was having exactly the results the NDP had predicted. The U.S. recession, which began in that same year, had stimulated an American repatriation of jobs, one of the consequences of having a foreign controlled economy. Canadian branch plants of U.S. companies, some of them the employment mainstays of their communities, were closing almost daily. North American motor vehicle production had fallen nearly 20 percent in the last year or so, hitting Ontario's auto industry-dependent economy hard. The effects of this loss of manufacturing jobs were cascading into the service sector and local economies, where property tax revenues started dropping. As pessimism tightened its grip on the province, the real estate market collapsed and housing starts fell, as did the sales of home furnishings and appliances. Northern Ontario resource industries, already reeling from the U.S. recession, were hit again as recession took hold in Canada. Mining and smelting activity dropped sharply, as did the forest products industry. All of this, of course, had a heavy impact on provincial tax revenues. During the election, the Liberals insisted they had a balanced budget. We learned immediately upon taking office this was not true. Every Cabinet meeting brought more bad news.

Bob Rae, Ontario's first NDP Premier, took office in late 1990 as the worst economic downturn since the Great Depression was hitting the province full force. Huge Hydro rate hikes were threatening industry and the Hydro debt was out of control. By 1995, rates had stabilized, the debt was dropping like a stone, a nuclear moratorium was in place and government-sponsored energy efficiency programs were measurably helping the environment. *Ontario NDP*

Not satisfied with killing a large chunk of our industry through the Free Trade Agreement, the Mulroney government compounded the misery by slashing unemployment insurance benefits, making deep cuts in transfer payments for education, health care and infrastructure programs. Ontario was particularly targeted for these debt-shifting strategies as Mulroney had already written off our province in the next federal election. Provinces more likely to stay with the Conservatives were given the money that had been withheld from Ontario. As if all this weren't bad enough, high interest rates coming out of Ottawa further strangled housing starts, car sales, small business credit lines and industrial investment. And how about the timing of the Goods and Services Tax (GST), the Conservative government's 1990 Christmas present to the nation? It was a bad, bad time to be in government, especially in Canada's industrial heartland.

As we struggled to keep the provincial economy from sinking into a full-blown depression, the Harris Conservatives had not a syllable of blame for their cousins in Ottawa. Instead, they ping-ponged between accusing the NDP of doing nothing to help the province and criticizing us for budget deficits that were essential to keeping public services from collapsing when they were needed most. The Liberals, now led by a former Minister of Energy, Lyn McLeod, followed suit. After all, none of this had anything to do with their nearly six years in power.

We especially took a lot of heat from the Conservatives and the right-wing press for raising social assistance rates in that cold winter of 1991, to try to take some edge off the pain of over 250,000 lost jobs, with more to come. I am personally very proud that our government had the political courage to do this in the face of the biggest economic downturn since the 1930s. The consequences of economic despair are well-documented: spikes in crime, family abuse and marital breakdown, suicide, children going hungry. To this very day, Ernie Eves tries to score retroactive political points by castigating the NDP government for the recession-era deficits. (Conventional economic history maintains that one of the reasons the 1930s Depression got out of hand was U.S. President Hoover's reluctance to run budget deficits at the time.)

The bad news from Ontario Hydro headquarters was equally distressing. We had spent a decade and a half trying to stop Darlington and now, in the very year that we formed our first government, it was finally completed, all $14.3 billion of it. And just as Darlington's first reactor started up, electricity consumption in Ontario began to drop for the first time since the early years of the Depression. We didn't need the new station before, and we clearly didn't need it now. Even worse, Ontario power consumers had to start paying for Darlington just as the recession hit. The law at that time stipulated that hydro rates could not include the cost of building a generating station until it began delivering power. This is not necessarily a bad idea, but the rate shock from Darlington could not have come at a worse time. From 1990 to 1993, as the Darlington units started delivering power, hydro rates went up more than 35 percent. Most of the increase was to begin paying for Darlington. Even these rate hikes were not enough to keep Ontario Hydro's debt from rising. You can't pay off $14.3 billion in three years. Naturally, the Conservatives and Liberals laughingly blamed the NDP for the Hydro debt increases that had to be shown on the books from 1990 to 1993, even though they had been caused by their own previous governments. You can hardly blame them for wanting to disown responsibility for this "mistake on the lake," but it was plainly dishonest to blame the consequences of their decision on the only people who had tried to stop it. We allowed ourselves a few minutes of griping about the unfairness of it all, but quickly sat down to figure out what to do. We couldn't change the past, but we could do something about the future.

The first order of business was to rein in Ontario Hydro and assert the government control that had been so badly lacking in the past. Ontario was awash in power, with a surplus of over 50 percent. Still, incredibly, Hydro claimed we needed more! For many years, the NDP had been pushing the idea of small-scale and diversified generation to reduce our reliance on coal and nuclear. Hydro's idea of small and diversified was an armada of 500-megawatt nuclear and coal stations dotted around the province. Their latest 25-year plan, unveiled a few months before the election, called for at least $50 billion worth of such stations. We all knew what Hydro's cost projections were worth.

We were determined to move Ontario Hydro's systems planning mindset away from its single-minded aversion to anything but mega-supply options. We knew this had to begin at the top so we worked with Hydro's Board of Directors to start changing the corporation's management culture. We had campaigned on a nuclear moratorium and quickly made good on this promise. In early 1991, we told Hydro to stop spending $240 million on pre-engineering work for three nuclear plants. The 25-year plan was undergoing a lengthy environmental assessment anyway, so even if we had supported nuclear, this would have made sense. Now, these stations will never be built. The NDP is the only Ontario political party to stop nuclear station development in its active planning stages.

We also began moving towards the small-scale, diversified generation mix we knew made more economic and environmental sense. We could not do this in an aggressive way because we had lots of power already, we were responsible for managing the huge Hydro debt we inherited, and we had no money. Some of our environment-focussed supporters wanted us to start dotting the landscape with windmills, solar houses and so on, but we said no, this made no economic sense in the middle of a recession with a huge overcapacity. They were for the next round. We had to lay the foundation for a more rational and environmentally friendly energy future, but we had to move slowly for now. We gave the green light to a handful of natural gas cogeneration, biomass and small hydro stations that were already in the planning stages, in order to gain experience with such projects. Even though Ontario didn't need the power, the size of the projects we approved meant that there would be no meaningful impact on rates.

We were saddled with a generation mix we didn't like and couldn't afford to start replacing, not for quite a while, anyway. We had one tool at our disposal, however, which had many benefits and would cost taxpayers so little money that it seemed almost magical: energy efficiency. I know that to most people, energy efficiency sounds rather mundane, hardly magical. But the economic and environmental benefits of energy efficiency are so great it is hard to understand why every government does not make it a public policy priority.

I can easily appreciate why *some* people are not interested in energy efficiency because they benefit from more power being consumed: deregulated utilities, power marketing firms, manufacturers of electric system equipment, uranium and coal mine owners, oil and gas companies, and developers of carelessly built housing. The rest of us, however, including all our fellow creatures on this planet, clearly benefit from lower energy consumption.

Energy efficiency is not exactly the same thing as energy conservation, though the terms are often used interchangeably. For me, conservation is when you turn down the thermostat so that you use less energy to heat your home. Energy efficiency is when you make changes to your home, such as putting in new windows and insulating your roof, so that you use less energy without having to turn down the thermostat. Same warmth and comfort, less energy. Another way of saying this is that the energy you use is more productive.

A government can encourage conservation, and should. You *should* turn out the lights when they're not needed. You *should* adjust the thermostat when you're not home. You'll save money and reduce environmental stress. But conservation alone as an energy saving strategy has limited usefulness. Schools and hospitals must be warm and well lit. Factories, mines and mills cannot turn their power down like a light on a dimmer switch and still maintain production. Computer-dependent industries require 100 percent-reliable power, with no voltage fluctuations. Cities must light their streets at night. A home should feel like a home, not a cave. Asking people to freeze in the dark while eating tins of cold beans is neither desirable nor sustainable public policy. Conservation is a good thing, but it is not the answer to fundamental concerns of supply, demand and environmental impact. Energy efficiency *is* the answer, or at least the lion's share of the answer.

We knew that by taking measures to get more productivity out of the energy Ontario consumes, we would lower the province's energy bill, freeing up money that could then be spent on other things. We would also reduce energy-related environmental impacts; create good jobs in the engineering and building trades, as well as in the manufacturing of energy efficiency technologies; and moderate power demand growth so that we could get off this treadmill of having to build a new nuclear or coal plant every few months, which is more or less what Hydro had been proposing for the past 25 years. Of all these objectives, job growth was uppermost on our minds at the time. The other benefits we knew would follow automatically.

We started in our own house. Shortly after taking office, Energy Minister Jenny Carter began a program of improving energy efficiency in the more than 7,000 buildings owned by the provincial government. It was a huge undertaking that would take several years to complete, so why not get started right away? It also made tremendous sense.

Retrofitting existing buildings to make them more energy efficient is virtually free because the costs of each project are paid for over time from the energy savings achieved. Imagine, for example, a large government office building with gas and hydro bills of $1 million a year. We might spend a million dollars to bring the building's lights, elevator motors, heating and cooling systems, and so on, up to current technological standards. The energy savings can sometimes be quite dramatic. Imagine cutting your lighting bill in half and getting better quality lighting in the bargain. Adding up all the improvements, our post-retrofit energy bill might drop by, say, $200,000 a year. This means that the project would pay for itself in five years, plus any financing costs. After that, we would keep all the savings in perpetuity, to the benefit of taxpayers. Even Buckingham Palace underwent a major retrofit in the mid-1990s, reducing energy consumption by about 25 percent. If it's good enough for the Queen...

We also prevailed upon Ontario Hydro to seriously ramp up its own anaemic energy efficiency efforts. In response, Hydro came out with a blizzard of small scale programs, such as rebates for compact fluorescent light bulbs and discount coupons for small energy-saving products you could pick up at the hardware store. All this was the prelude to an early 1992 announcement by Hydro of a master plan to reduce provincial power demand over the next eight years by 5,200 megawatts.

That's a lot. At the time, it was about 25 percent of the province's peak demand and represented 1.5 times the amount of power produced by Darlington, exactly what had been called for by the NDP years earlier. Over $6 billion would be invested and 50,000 years of employment created. It was typical Hydro grandiosity, but this time it was appropriate. The money and employment would necessarily have to be spread all around the province. It wasn't Hydro employees who would be doing the retrofitting work, it would be local heating contractors, electricians, glaziers and other trades. Moreover, even if Hydro "pulled a Hydro" on us and the program ended up costing twice as much, that would still be about half as much as Darlington cost, per megawatt. As a bonus, there would be no radioactive wastes to deal with. And how's this for irony: one of the main features of this program was to encourage "fuel switching," which meant changing from electricity to natural gas for heating and other applications. After 25 years of building nuclear plants in order to fulfill energy needs better served by natural gas, Ontario Hydro was now *paying* its customers to convert to gas. Well, better late than never, I guess.

Tragically, this ambitious plan was sidelined once the Conservatives took control in late 1995. Since the Conservatives intended to privatize Ontario Hydro, reducing provincial power demand was just not on the agenda. A private generator can't make money on power it doesn't sell. Moreover, as consumers in all deregulated jurisdictions, including Ontario, have learned so well, the higher the demand, the higher the price. Private power companies are as interested in energy efficiency as Exxon is in bicycles. Had Ontario Hydro continued the program begun under the Rae government, we would not have had the power shortages we are now experiencing and we would not now be choking as badly on (and dying from) the smog created by our coal plants.

As kids, most of us were admonished to "Close the refrigerator door!" For good reason: the always-on refrigerator is a big consumer of electricity. In homes with gas furnaces and water heaters, the refrigerator may well be the single largest power user. This is why we set new standards for refrigerator efficiency shortly after coming to power. By 1994, all new refrigerators sold in Ontario would have to be twice as efficient as the ones then being sold in early 1991. This is another one of those mundane-sounding things that is really quite significant when you examine, as you must, the cumulative effect. At the time, the typical Canadian-made refrigerators used between 110-140 kilowatt hours per month. We knew that with existing technology that could be more than cut in half, easily. In fact, an appliance manufacturer in Guelph was making refrigerators that used only 36 kWh, one-third the energy needed for typical models, and was selling them for about the same price as the inefficient ones. Ironically, most of these refrigerators were sold into the American market.

"Okay, so big deal," you say, "I save, at the very best, 100 kWh per month. What's that worth?" At current "all-in" average hydro prices, that's around $8 a month, almost $100 a year. At some point the cost of your refrigerator will be paid for out of the energy savings, likely long before its useful life is over. Although it's not a huge deal, the energy savings do pay for the fridge, the same principle as other retrofit measures. Now multiply your home's energy savings by the four million or so residential-sized refrigerators in the province and, voilà, we need one less Pickering-sized nuclear reactor. Little things can add up to big things.

We did hundreds of other "little things" during our years in government to reduce energy use, conserve water, curb pollution and generally make Ontario a greener place. A few examples: we followed up the refrigerator initiative with several more like it, covering 26 product categories; we worked with financial institutions like Canada Trust to make millions of dollars of non-government money available for home energy retrofits; we assisted industries that wanted to improve their power and water use—a gratifying number were interested; the Council on Renewable Energy was established; First Nations communities were financially encouraged to pursue small-scale green power developments; and hundreds of thousands of homes were visited and residents advised on inexpensive energy and water conservation measures under our popular Green Communities program.

These and other NDP environmental initiatives were held in great disdain by the Conservatives. Many of our programs were summarily executed by the Conservatives in 1995. Here's something you should know: NDP government spending on environmental protection averaged nearly $700 million per year, almost 25 percent higher than under the previous Liberal government. The Conservatives, during the most sustained economic boom in North America in generations, cut environmental spending *in half*. The deaths in Walkerton from E. coli-poisoned water in 2000 are only the most high-profile results of the Conservatives' "common sense." I shudder to think of how much worse the killer water situation could have been had our government not already found and corrected significant compliance problems in 25 percent of the province's nearly 500 water treatment plants. Inadequate testing for bacteria and toxic chemicals was alarmingly common. We acted swiftly to correct this public health hazard by setting more stringent testing standards and requiring operators of water treatment plants and sewage treatment plants to be licensed. More red tape, as the Conservatives would say. We also beefed up enforcement. The Harris government downsized inspection.

The contempt the Conservatives seem to have for Mother Nature and her defenders is as wrong economically as it is morally. The American biologist Edward O. Wilson, one of the world's most influential scientists, has estimated the economic value of services provided to humanity free of charge by the living natural environment to be at least US $33 *trillion* annually, double the GNP of all countries of the world combined. The gift of clean water from forest-covered watersheds alone is worth hundreds of billions of dollars to North Americans. That's how much we'd have to spend on water purification plants to do the same job. If we neglect our environment, Wilson says, we eventually kill ourselves, but not without first spending a lot of money to replace the utility of what has been destroyed through our ignorance.

This is why, in my view, one of our government's most important achievements was the Environmental Bill of Rights. Among several other ground-breaking provisions, it gave Ontario citizens the right to bring legal action against polluters and also protected environmental "whistle-blowing" employees from reprisals by their polluting employers. My Cabinet colleague Ruth Grier, did a magnificent job of orchestrating a broad public consultation on the proposed bill throughout 1992, forging a consensus among environmental, labour and business groups that many thought impossible. In early 1993, Premier Rae combined the Energy and Environment portfolios and appointed my fellow northerner Bud Wildman as Minister, moving Ruth to oversee Health. Bud picked up the project with great enthusiasm, shepherding it carefully through the Legislature and extensive public hearings. Ontario's Environmental Bill of Rights became law in 1994. I urge you to visit the website of the Environmental Commissioner of Ontario, a permanent office that was created by our law.

If I were to list my own environmental contributions during the Rae government, certainly near the top would be my role in developing the Forest Renewal Trust Fund and the Crown Forest Sustainability Act, both revolutionary changes in how we view our public forests. They are the culmination of decades of work and dreams of New Democrats, particularly my northern colleagues.

With the Forest Renewal Trust Fund, stumpage fees paid by forest industry companies were dedicated to ensuring the renewal of the forest for future generations. The industry took direct responsibility, for the first time ever, for guaranteeing that every tree removed would be replaced. Money generated from northern forests would return to northern forests instead of being funnelled into general revenues and being subject to being scooped up by the government and the political priorities of the day. This idea was anathema to Treasury bureaucrats in Toronto, but was a breath of fresh air for residents of northern communities like mine, who had always seen money flow south, but rarely north.

The Crown Forest Sustainability Act went hand-in-hand with the Forest Renewal Trust Fund. Previously, thanks to the politically powerful timber barons, when governments viewed the large tracts of Crown land, they saw newsprint, chips and sawlogs and large dollar signs. The CFSA changed all that, and its significance cannot be overestimated. For the first time, all the values of the forest were recognized and given weight. Communities that depend on forest industry jobs, tourist resort owners, anglers, naturalists, hunters, canoeists, all have an interest in how the forest is managed and those interests were finally protected. Above all, the prior interest of our First Nations had to be respected and honoured.

I am still very proud of this accomplishment, even after a number of years. The Crown Forest Sustainability Act and the Forest Renewal Trust Fund never received a lot of media coverage in Toronto, but the changes they brought were profound, progressive and have endured even as the Conservatives have tried to turn back the clock.

In all of our efforts to improve Ontario's environment, we preferred to work hand-in-hand with local governments and hydro commissions, as well as community groups, including (some will be surprised to learn) local business associations, pulp and paper companies, mining companies and chambers of commerce. There is a myth, gleefully promoted by Conservatives and Liberals, that the NDP thinks "big government" is the answer to all of society's problems. This is patently false, as our record shows. Our world view sees neighbourhoods and communities as often the most efficient and accurate way to pinpoint economic and social needs and then deliver a solution. As for the related myth that the NDP is anti-business, I would gladly put our record of consulting with Ontario businesses against those of the Liberals and Conservatives. It was a rule in the Rae government that we consult with recognized business representatives on *everything* we were planning to do that would affect them. We also made sure that we didn't just hear from the corporate heavyweights, taking great pains to ensure small and medium-sized businesses had a voice in the future. We found, in this way, that smaller businesses provide the most acutely sensitive barometers of Ontario's economic health and direction. In addition, they are more operationally flexible and responsive to change. Our home retrofit programs, for example, could only be effectively delivered by smaller firms. Of course not all solutions are local and small-scale, but many are. Like more efficient refrigerators.

We also did big things. In 1992, Premier Rae persuaded Maurice Strong to take the helm of the floundering Ontario Hydro. The appointment gained worldwide attention. Strong was a successful Canadian businessman who had been a senior advisor to the Secretary-General of the United Nations. He had recently gained global recognition as the organizer and Chair of the 1992 Earth Summit in Rio de Janiero, the largest international environmental conclave in history. The Summit produced, among other breakthroughs in environmental cooperation, the United Nations Framework Convention on Climate Change, which ultimately led to the 1998 Kyoto Accord. There was talk of Strong's becoming the next U.N. Secretary General. Fortunately for us, he agreed to take on the formidable challenge of rescuing and reshaping North America's largest utility.

He didn't waste any time. In early 1993, Strong stopped Hydro's still-rampant construction program in its tracks. Planned projects worth $24 billion were cancelled, many that had already been started. It is true that as a result, 8,000 jobs were lost, but we were determined to set a new energy course. The Power Workers' Union understandably complained, but almost all the job terminations were voluntary. Hydro's good pension plan, in place since the 1930s, was able to handle a surge of early retirements and the rest received good severance deals and other assistance. Admittedly, the restructuring costs were heavy. Hydro reported an operating loss of over $3 billion that year, but it was the last loss at Hydro until the Conservatives took over. Here's a fact: the cumulative losses at Hydro in 1996-97, under the Tories' Bill Farlinger, were $8.3 billion, something Ernie Eves neglects to mention when he drones on and on, as he loves to do, about "the $3 billion that the NDP put on the Hydro credit card."

The 1993 hydro rates had already been settled by the time Strong arrived. At our government's direction, he froze them for the following two years. The persistent recession was causing even more of Hydro's industrial customers to cut back production or even shut down. Cash flow was extremely tight and many smaller manufacturing operations were skirting the edge of bankruptcy; some were slipping over the edge. Many of these industries were big energy users. If we could pre-

Maurice Strong, left, and Amory Lovins. As the NDP-appointed chair of Ontario Hydro (1992-95), Strong transformed the giant utility from a money-losing albatross intent on building more nuclear and coal stations, to a profitable public asset committed to sustainable development. Strong often consulted with Lovins, the world-renowned energy efficiency expert and head of Colorado's Rocky Mountain Institute. Lovins coined the term "negawatts" to quantify the amount of generation capacity not needed as a result of energy efficiency measures. *Alex MacDonald*

vent some of them from bankruptcy by freezing one of their largest input costs, we would save jobs, protect provincial tax revenues and even, in some cases, save whole communities. Had it not been for Darlington, we might have even been able to roll back hydro rates for a couple of years, at least for industrial customers, without damaging the province's credit rating. In fact, we later did that for a number of large energy-intensive industrial customers, using our huge power surplus to help keep them competitive.

To our delight, though not surprise, Maurice Strong turned Ontario Hydro around in what was, in bureaucratic terms, the speed of light. New senior management was brought in, such as John Fox, an Ontario native who earned a reputation as an aggressive promoter of utility-driven energy efficiency programs in the U.S., particularly in California. Many were from the private sector. All reflected our determination to make Hydro politically accountable and environmentally sensitive. More energy efficiency programs were launched. A five-year, $110 million commitment to renewable energy technologies was made. An accelerated settlement of First Nations grievances was undertaken. Developments in the U.S. pointing towards deregulation were carefully assessed for their potential impact on us. An internal system of price and reliability "competition" was introduced among Hydro's 80 generating stations, to reward performance improvement. A greenhouse gas emissions management strategy was adopted that exceeded Canada's commitments under the Climate Change Agreement Strong had brokered in Rio. The world's first facility for the testing of pressurized solid oxide fuel cells was opened by Ontario Hydro Technologies.

And the Ontario Hydro debt? In 1993, it went down for the first time in many years. We now owed about $400 million less on the Hydro "mortgage." The following year, 1994, it dropped by another $700 million. In 1995, during Maurice Strong's last year before being replaced by the Conservatives in November with Bill Farlinger, the Hydro debt declined a further $1.5 billion. That's a total of $2.6 billion off the debt in the last three NDP years. At the end of 1995, it stood at $31.4 billion. These figures are all a matter of public record. So I find it odd that Premier Harris and Premier Eves and their various energy ministers have always been so fond of citing the "$38 billion Hydro debt" as proof that our publicly owned system should be broken up into smaller lots and sold off like so much junk hauled from the attic.

Premier Rae called the spring election on April 28, 1995. It would be a six-week campaign. All parties had been gearing up for a year. Inside the NDP, optimism was in short supply. We were well down in the polls, despite an increasingly visible economic turnaround. All important indicators were good, very good in fact. Ontario led Canada in growth in 1993 and 1994. The auto industry had committed to billions of dollars in new investments. Housing starts were up. The north was in better shape than it had been in decades. Mills we had helped save from going under during the recession were now working around the clock to meet burgeoning U.S. demand. Ontario Hydro had turned around on a dime. This good news, however, had not yet reached the voters. There is typically a lag time between economic truths and political consequences. As luck would have it, we were obliged to call an election before the recovery became common knowledge. We had other problems as well, such as the controversial Social Contract that gave public sector workers a few unpaid days off—Rae Days—to help reduce the provincial deficit. We thought a six-week campaign would give our candidates and our thousands of still-loyal volunteers the chance to spread the good news and get people to think about the consequences of turning back the clock.

Mike Harris and the Conservatives made poor people the main issue, demonizing women and children who had to rely on social assistance as responsible for our allegedly intolerable tax burden that was supposedly driving away business from the province, which would ultimately impoverish us all. It was a sickening display of political cynicism that I pray I never witness again.

Ontario Hydro was not a big topic in the election. Both Harris and Lyn McLeod, the Liberal leader, mused about privatization and competition. But with hydro rates under control, the Hydro debt coming down, and no new fiascos to focus on, most of the electorate were unconcerned. "Is it there when I need it and can I afford it?" is what most voters want to know about Hydro. A growing number also want to know: "What impact is this having on the environment?" Although Hydro was not a major election issue in Ontario in the spring of 1995, there was much going on, in the United States and elsewhere, that would soon move it to the front burner of the political stove.

12

The British Disaster

Privatization and deregulation in the United Kingdom's power sector were meant to fix its inefficiencies. The purported cure turned out much worse than the alleged disease.

What about privatization and deregulation? Where did these ideas come from? Where have they been adopted? What has been the experience elsewhere and what can we learn from it?

The generic ideas of privatization and deregulation have been around for as long as their implied opposites, public ownership and regulation. Indeed, much of human history can be viewed as one long running argument about the role of government in the economic life of society. In the case of the electric power industry, this argument has been going on non-stop since the industry was born in the 1880s and will still be in progress, we can be assured, in the 2080s. Over the past two decades, the forces of electric utility privatization and deregulation have become something of a fad, a fad that needs critical review, particularly in light of recent experience.

The first country to restructure its utility industry through privatization and deregulation was not the United Kingdom, as many think, but Chile, that long, thin strip of a country on the Pacific side of South America. After the democratically elected government of Salvador Allende was overthrown by an American-supported military coup in 1973, Chile became the intellectual playground of American conservative economists like Milton Friedman, who believed that the less regulation, the better. Under the personal tutelage of Friedman, the military dictatorship of General Augusto Pinochet embarked on the world's first broad-spectrum program of privatization and deregulation. Utility restructuring legislation was put in place in 1982 and implemented over the next few years. Two major features of that restructuring were widely adopted by those who followed Chile's lead: the functional segmentation of the industry into the generation and wires sectors, and the compulsory spot market, which meant that all electricity was sold through a centralized daily bidding system.

I do not think we in Canada need to take economic lessons from a military dictatorship, thankfully now gone. But former British Prime Minister Margaret Thatcher, a keen admirer of Friedman, had no such qualms and adopted much of the Chilean model. And since the U.K. is a vigorous and stable parliamentary democracy, like our own, we can and should pay attention to its experience when shaping the future of our own electricity system. Certainly the Ontario Conservatives and Liberals have done so. Both of them have often expressed admiration for the way the U.K. privatized and deregulated its electricity system. "If they can do it, so can we" has been their simplistic notion.

On June 11, 1987, Margaret Thatcher became the first British Prime Minister in over a century to win three successive election victories. The Conservative party platform for that election had promised to break up, privatize and deregulate (in Europe, they call it "liberalize") the country's electricity industry. No one doubted Thatcher would fulfill this campaign commitment. As she herself famously said, "This lady is not for turning."

As in most of the Western world in the late 1980s, Britain's economy was booming. The country was in the most sustained period of economic growth since the Victorian era. Since her second election in 1983, Thatcher had privatized many of the industries that had been nationalized in the aftermath of World War II. British Telecom, British Steel, British Gas, Rolls Royce, British Airways, London Transport and the water and sewage companies had all been divested. Many of these public enterprises were profitable and performed well in public hands, but that didn't matter. Private was good; public was bad. By the time Thatcher was ousted in a palace coup in late 1990, about 40 public sector companies had been privatized. A number of these newly privatized companies did well at first. How much of that success was due to privatization and how much to the swelling tide of prosperity in Western Europe and elsewhere is still debated. But to the voters who were bringing home bigger paycheques and, in many cases, cashing dividend cheques as well, doing the same thing to electricity seemed like another of the Iron Lady's good ideas. Two years later, in July 1989, the Electricity Act passed Parliament and the grand experiment began, setting the basic pattern for power industry restructuring throughout much of the English-speaking world. Few knew that the military dictatorship in Chile provided the original model.

Britain's electricity system prior to its privatization looked much like Ontario's. A Central Electricity Generating Board controlled 95 percent of all generation and all of the country's transmission — their version of Ontario Hydro. The distribution and retail sale of power to end users was handled by 12 Area Boards, analogous to our municipal electrical utilities, albeit larger. The Thatcher government's original idea for restructuring was to break up the generation industry into two entities and then sell them off: National Power, which held 70 percent of the country's power capacity, including all its nuclear stations, and Powergen, which held the remaining 30 percent. It didn't work out quite like that, however. The nuclear stations were deemed by the financial markets in Britain to be "unfit for public consumption" and had to be withdrawn from the sale. The nuclear facilities were carved out of National Power and reconstituted as Nuclear Electric Ltd., which continued to be wholly state owned for a few more years. Sixty percent of the shares in National Power and PowerGen were then sold in March 1991. The rest were sold in 1994. The Area Boards were renamed Regional Electricity Companies (RECs) and also privatized in 1991. The RECs initially had control of the transmission system in their areas, but a few years later all transmission assets were consolidated into one company, National Grid, which was sold in late 1994. All wholesale power was sold through the Friedman-inspired Electricity Pool, which was similar in many respects to the system now run by Ontario's Independent Electricity Market Operator.

Competition was at first restricted to large customers whose demand exceeded one megawatt. This included the RECs, which were supposed to pass on the savings they realized from wholesale competition to their smaller, still captive customers. A regulator was put in place to ensure that the RECs did not abuse their lingering monopolies and that the Electricity Pool was truly competitive. In 1994, competition was extended to customers with demands over 100 kilowatts, which would include most medium-sized businesses. In 1998, all customers were to be put into the so-called competitive market. They could either stay with their REC or move to another provider. In 1996, Nuclear Electric merged with Scottish Nuclear to form British Energy, which was promptly privatized. The initial sale of shares netted barely half what had been expected by the Conservative government of John Major, Thatcher's successor. This was acutely disappointing as the money was to be used for 1997 election year goodies, in the same way that proceeds from previous privatizations had been used in the last two elections. The sale went through nonetheless.

Was the Conservative privatization program a success? If you were one of the new companies' shareholders, most definitely. In the initial round of privatization of the RECs, National Power, Powergen and National Grid, shares were sold at what proved to be well below their market value, a bonanza for those who got in first. This was especially the case for the utilities' managers, who were given generous stock options along with big pay increases. They quickly became known as the utility "Fat Cats," who effortlessly made millions from the Conservatives' giveaway, joining the fat cats created in prior privatizations of other publicly owned enterprises. Meanwhile, many thousands of utility workers lost their jobs and every time a large layoff was announced, share prices rose. This enrichment of a relative handful, clearly at the expense of workers and taxpayers, was the least popular feature of U.K. privatization. But the shares continued to increase in value as the profits of the private utilities soared throughout the 1990s.

Flexing their financial muscle, the new companies went global. National Power and Powergen began buying and building generating stations in ten countries around the world. National Grid now owns extensive transmission and distribution assets in the U.S. Northeast. The sale of the New England Electrical System and Massachusetts Electric to National Grid was not warmly received by New Englanders, who still like to brag about how they defeated the British Army over 200 years ago. The famous call to arms: "The British are coming!" was heard once again, but in the end money, not patriotism, prevailed. Scottish Power went way out west to purchase Oregon's Pacificorp, which serves 1.5 million customers. U.S. consumer advocate Ralph Nader tried unsuccessfully to stop this controversial takeover. British Energy joint-ventured with U.S. utilities to purchase, for a pittance, a handful of troubled Pennsylvania nuclear reactors, including Three Mile Island, site of North America's worst nuclear accident in 1979. British Energy also became for a while the lion's shareholder in Ontario's Bruce Power, which controls the world's largest nuclear plant at Kincardine, on Lake Huron.

All these offshore investments were, in effect, financed by U.K. electricity customers, whose inflation-adjusted power prices have declined between 1990 and 1999, but almost entirely as a result of the regulator ordering price cuts. Least to benefit were the country's 26 million residential customers. Taking inflation into account, their bills dropped about 15 percent between 1993, when power prices started to come down, and 1999. This was only marginally better than the inflation adjusted power price reductions of about 11.1 percent for Ontario customers, whose rates were frozen during the same period by the provincial government, first the NDP and then the Conservatives. The fact that U.K. customers were getting the crumbs left over from the shareholders' feast led to considerable discontent towards the end of the Conservative era. A Labour Party promise to steeply tax the runaway "windfall" profits of the privatized companies was a popular plank in its 1997 election platform.

The environment has been a big winner in Britain, not as a result of privatization, but rather as a direct result of the widespread adoption of new generating technologies that could exploit the vast reserves of newly discovered North Sea natural gas. Utilizing North Sea natural gas to produce power cut the share of power generated from coal in half. In 1990, coal accounted for 70 percent of all generation in England and Wales. Today, it's about 35 percent. The proliferation of new natural gas-fired plants, which are not only much cleaner and quicker to construct, but also much more efficient than older coal plants, is the single largest reason for almost all of the decline in U.K. power prices since privatization, as well as the steady growth in private power utility profits throughout the 1990s.

This reduction in coal generation was not a result of privatization. The discovery of cheap North Sea natural gas would have been available to a publicly owned system. Private power advocates might claim that the British "dash for gas" happened more quickly than it would have under public power, but that would be speculative nitpicking. The fact is that the entire utility world, including the publicly owned sector, has been installing combined cycle gas turbine (CCGT) generation for the last 20 years or more. It simply makes economic and environmental sense to replace coal with natural gas, assuming you have secure access to good supplies, which the North Sea provides. You don't have to be a highly paid "fat cat" with lucrative stock options to figure this out. You could do that by skimming through back issues of *Popular Mechanics*. Let's not credit privatization with what would have happened anyway.

While British electricity companies were fanning out across the globe, foreign investment money flowed in. A number of the RECs were sold to foreign interests. Enron began building gas plants, as did another Texas-based marketer, TXU Energy. So did American Electric Power, Edison Mission Energy, the Southern Company and Calpine, among other U.S. companies. The publicly owned French utility, Electricité de France (EDF) bought the country's largest distributor, London

Electricity. In 2002, German utility giant E.ON bought 100 percent of Powergen. Billions and billions of dollars moved around the globe as utilities and marketers in many different countries took in each other's laundry, so to speak.

On May 1, 1997, Tony Blair led Labour to its biggest ever general election victory. The promised windfall profits tax on privatized utilities was implemented and power price restraints were put in place. The Labour government postponed full retail competition of electricity and gas suppliers until 1999. This final step towards competition touched off a plague of fraudulent marketing tactics called, with characteristic British understatement, "mis-selling," which means tricking people into changing suppliers and not explaining the financial consequences (in the more aggressive U.S. culture, such practices are called "slamming"). While mis-selling is not the problem it once was, it's still around. A profit-driven market inevitably attracts cheaters, as we have found out here in Ontario.

In 1998, the U.K. utility shareholder nirvana began coming to an end. The Energy Intensive Users Group called for an investigation into market manipulation by the large generators. By some remarkable coincidence, Electricity Pool prices immediately dropped 29 percent. In January 2000, the first Labour-appointed regulator, Callum McCarthy, accused generators of manipulating the wholesale power market—the Electricity Pool—to keep prices high. In a strongly worded high-profile speech, McCarthy said the generators were exploiting "fundamental flaws" in the rules for "unacceptable" commercial gain. Gaming the Pool rules had cost customers over $200 million (Cdn) in December 1999 alone. Pool prices remained "stubbornly high" and had, he said, increased since privatization, despite sharp falls in input prices over the past decade. He vowed to use his statutory powers to stop the generators' abuse of the market. Utility share prices fell. Finally, in March 2001, the Labour government scrapped the Friedman-inspired Electricity Pool and established a different market framework called New Electricity Trading Arrangements (NETA). NETA allowed for bilateral contracts and futures trading. Large customers, RECs and mid-level marketers could now negotiate wholesale prices directly with individual generators rather than being forced to buy from a pool that could be manipulated with relative ease. Retail customers can stay with their REC or buy on contract from a mid-level marketer. Centrica, which sells both gas and electricity, is the largest such marketer in the U.K. Centrica owns Direct Energy, which markets gas and electricity here in Ontario.

The new regulatory model has had a major impact on prices. As of October 2002, U.K. wholesale prices averaged 40 percent less than in 1998. The main reason is that customers are now able to negotiate more of the industry's input cost reductions, especially those attributable to the newer natural gas plants, of which there was an "oversupply." The Labour government's new regulation of the U.K. electricity market has, in a nutshell, forced the generating companies to turn over to consumers more of the power cost reductions that came from cheap North Sea natural gas.

The U.K. market began "working" so well that British Energy, which supplies 20 percent of all the power in Britain, was shipwrecked and forced to get a government loan of nearly $4 billion (Cdn) just to stay afloat. Last year it lost $1.2 billion. Following the introduction of NETA, wholesale prices fell below British Energy's cost of production, about $46 per megawatt hour. For natural gas, the cost of production is $10 less. British Energy continues to produce and sell power anyway because you can't just turn a nuclear reactor on and off in response to fluctuating market conditions. Besides, even if the company loses money producing power, it would lose even more by not producing power at all. There's still a huge nuclear debt to pay down. The Labour government, distressed at this backfiring of "at long last, true competition!" had no choice but to step in with public money. Allowing the nuclear plants to abruptly stop operating could trigger an economic crisis. It would certainly lead to energy shortages and, naturally, much higher prices. Moreover, much of the lost nuclear power would have to be made up by de-mothballed, uneconomic and polluting fossil fuel plants. This would be a huge environmental blow and would preclude the U.K. from meeting its current international emissions reduction commitments.

British Energy's future is bleak, operating nuclear generating stations that will never be able to compete head-to-head with natural gas in an open market. It's not really a matter of making British Energy more efficient. It's just a fact that nuclear energy has unavoidable costs that seem to grow over time. Operational cutbacks can be hazardous, as Ontario Power Generation knows. There are rather strict limits to how much you can trim costs in a nuclear station without compromising safety. You don't want to go down that road. According to deregulation and privatization theorists, you could perhaps slash wages and benefits and push the workers to do even more, for less, although that's another conservative economic theory that doesn't work in the real world. Even if you could run a nuclear generating station with unpaid robots, this still wouldn't bring nuclear prices down all that much. The fact of the matter is that power prices from reasonably well-run nuclear generating stations are fairly inelastic. Far from being too cheap to meter, there is a floor price that must be sustained if a nuclear generating station is to be run safely and without a deficit. As things stand now, U.K. taxpayers are stepping into the breach to prop up British Energy's losses. Advocates of deregulation will, of course, not take this taxpayer subsidy into account when they point to wholesale power price reductions under the Labour government and say: "See, it works over there." Nor will they volunteer that most of those price reductions have yet to flow to the large number of Britons who signed long-term fixed-price electricity supply contracts with mid-level marketers like Centrica, whose profits have zoom-zoomed since NETA.

British Energy is not the only U.K. generator in trouble. The German-controlled Powergen has mothballed 25 percent of its U.K. capacity in recent months. American-controlled TXU has pulled out of the market, selling out to Powergen. Several other generators are on the brink of shutdown. There's not enough money to be made. The market is working; it's punishing overcapacity. The solution is inevitable. The only way for prices to reach a sustainable (i.e., profitable) level is by taking generating capacity off-line, which is happening. Prices will go up again in response and the market will then, as always, be working. By the time those millions of locked-in retail customers finally get the chance to renew their power supply contracts, wholesale prices will be up again.

What does the U.K. have to show for more than ten years of privatization and deregulation? A bankrupt nuclear company kept from sinking by the taxpayers. A non-nuclear power generation sector that is largely foreign controlled. A number of foreign-controlled distribution companies. A transmission company that was built by public enterprise but is now owned by private shareholders who use their U.K.-generated profits to invest elsewhere. Lower wholesale prices, for a while at least, but not meaningfully lower retail prices. The U.K. marketplace is rife with uncertainty and small customers are forced to pay a premium to mid-level marketers like Centrica in order to stabilize their power rates and have some security, just like here in Ontario.

Not much of an improvement.

If you're tempted to think that the U.K. example is just a privatization anomaly, that perhaps it didn't work so well with electricity, but privatization is still fundamentally a sound economic policy, consider the case of Railtrack, the privatized British Rail. The once privately owned railways in Britain had been nationalized in 1947 and consolidated into a single integrated network. This idea had been around since at least 1918, when Winston Churchill remarked: "So long as the railways are in private hands, they may be used for immediate profit. In the hands of the state, however, it might be wise or expedient to run them at a loss if they developed industry, placed the trader in close contact with his market and stimulated development." When the post-war Atlee Labour government took the Conservative Churchill's advice, the existing shareholders received generous compensation, considering the dilapidated state of the system they handed over. Large amounts of public money went into upgrading British Rail; most of that money went to private sector contractors and suppliers, which stimulated the economy. Public subsidies continued in order to keep fares within reach of all and to slow down the trend towards private automobiles. Passenger-for-passenger, highways cost a lot more to build and maintain than railways.

The Conservatives began privatizing British Rail in the mid-1990s. The single integrated network was broken up into some 40 companies, 25 of which operated the passenger trains. Some of these companies operated just one line, a prescription for scheduling chaos. British Rail's freight service was sold to an American company for less than one-tenth its asset value. Public subsidies continued and even increased, putting the lie to the Conservatives' claims that privatization would benefit taxpayers. Maintenance suffered because of the inherent conflict between profits and expenses. In a few short years, the new operators turned the country's rail system into a disaster. Fares on some routes came down slightly, but service levels declined precipitously. Overcrowding became common. Passengers deserted and clogged the highways, forcing a massive increase in public investment in roads infrastructure. Several major rail accidents occurred after privatization, the last one in May 2000, killing seven people. "Deferred" maintenance was blamed. Shareholders, however, collected nearly three quarters of a billion dollars in dividends. In October 2001, the Labour government finally took over Railtrack and paid nearly $3 billion to the shareholders. A new government-backed debt-funded business has emerged called Network Rail. Effectively, the rail system has been re-nationalized. It is estimated that rebuilding the rail system will cost nearly $20 billion—an expensive lesson, like British Energy, in the perils of privatization.

13

The American Disaster—Part 1

Deregulation in the U.S. had its origins in global oil politics, expensive nuclear mistakes and federal government efforts to strip regulatory power from the states.

Electricity deregulation in the United States followed a very different path than in the United Kingdom. There are three reasons for this. First, privatization was not an issue in the U.S. Three-quarters of the industry was already investor-owned and no one was proposing selling off the publicly owned sector. Indeed, that would have made no economic sense. Throughout the U.S., publicly owned electric utilities offered significantly lower average power rates. Secondly, utility industry structure was very different in the two countries. In the U.K., the central government owned the entire electricity system and was its sole regulator. It could impose a uniform set of rules on a restructured industry. In the U.S., utility ownership was very diverse, with hundreds of generating companies and thousands of distributors. There were also 50 state public utilities commissions answerable to 50 state legislatures. Achieving one set of rules to govern the entire U.S. power industry was clearly impossible. Thirdly, regional power prices in the U.K. varied little, if at all. Competing factories in Birmingham and Manchester paid essentially the same for their power. One plant might use the power more efficiently than the other but there was, at least, no electricity input cost advantage based on geography. In the U.S., by contrast, regional prices varied considerably, and still do. In the early 1990s, manufacturers in Pennsylvania and New York, for example, had to pay twice as much for power as their competitors in Kentucky or Montana. It is therefore not surprising to learn that the main impetus for U.S. deregulation, at least in its early stages, came from energy-intensive industries in those states with above-average power prices.

While there had always been regional power price differences in the U.S., they were relatively small and unimportant prior to the 1970s. Oil and coal accounted for about 80 percent of all generation and both were cheap. Moreover, power prices had been steadily declining since the utility industry began, through a combination of technological advances and regulation that assured that the benefits of those advances were shared between producers and consumers. In 1967, average U.S. retail power rates were about nine cents per kilowatt hour. In inflation-adjusted dollars, that same kilowatt hour had cost nearly fifty times more in the 1890s, when Samuel Insull first began to amass his empire. Many thought that this historic trend of falling prices would continue until power rates approached zero. They proved to be wrong, for essentially the same reasons that similar predictions in Ontario turned out wrong.

In the 1950s and '60s, many American utilities had fallen into the same trap as Ontario Hydro. They promoted the use of electricity usage through artificially low rates and "Live Better Electrically" campaigns. As demand increased, more generating stations would be built. These newer stations would supposedly produce power at lower marginal cost, further bringing down rates. For private power companies, more generating capacity also meant automatically higher profits. Why? Even though rates were regulated, they always included guaranteed profits on invested capital. Build more generating stations, make more money. The incentive of additional profits makes the U.S. demand-building campaigns easier to understand than in Ontario's non-profit system. But, as in Ontario, the U.S. campaigns worked. Throughout the 1960s, consumption soared, appliance manufacturers cheered and real electricity rates continued to decline.

Suddenly, in October 1973, international energy economics changed forever. The world's longest-running territorial dispute, dating back to Biblical times, broke out into war once again, when Israel, Egypt and Syria clashed. As punishment for its longtime support of Israel, the Arab states-dominated Organization of Petroleum Exporting Countries (OPEC) stopped shipping oil to the United States. The embargo lasted five months. Since 36 percent of all U.S. oil was imported, mostly from the Middle East, the embargo traumatized the economy. About 20 percent of American electricity was generated by oil-fired plants. Oil-burning furnaces heated millions of homes. Then there were the millions of cars. Petroleum prices quadrupled overnight. It was America's first nation-wide energy crisis. It was also the best thing that ever happened to the nuclear and coal industries.

Within weeks of the embargo, President Richard Nixon put gas rationing measures in place. He also assured Americans that by the end of the decade the U.S. would be self-sufficient in energy. "Project Independence" would achieve this by stimulating domestic oil production (through higher prices), by converting oil generating stations to coal and by building as many nuclear plants as possible, as fast as possible. Part of the plan worked. In 1973, nuclear accounted for only four percent of U.S. electricity production because very high capital costs made it uncompetitive. By the 1990s, nuclear accounted for 20 percent of all U.S. power, oil barely three percent, a complete reversal in less than 20 years. Coal was still king, as it had always been. At its height in the mid-1990s, coal produced over 60 percent of all electricity in the country. The other part of Nixon's plan, reduced dependence on foreign oil, didn't work. More than half of all oil consumed today in the U.S. comes from foreign sources, much of it from the Middle East.

The U.S. energy crisis didn't get any better under President Jimmy Carter, despite his earnest attempts to get Americans to drive less, drive slower and turn down the thermostat. Once again, the politics of oil intervened. In 1979, when Iran's new revolutionary government allowed American embassy staff in Tehran to be taken hostage, it also cut off Iranian oil to the U.S. "Market forces" took advantage of the abrupt shortage and prices again skyrocketed. Carter addressed the nation in December and asked for "the most massive peacetime commitment of funds and resources in our nation's history to develop America's own alternative sources of fuel." Carter called the drive for energy independence "the moral equivalent of war." His critics, who simply wanted to bomb Iran and take its oil, belittled the plan as the "meow strategy." The previous year, however, Carter had already launched a long-term energy independence program by signing the Public Utility Regulatory Policies Act (PURPA). The purposes of this law were to promote fuel conservation through "cogeneration" and to encourage greater use of alternative sources of power generation. PURPA established a class of non-utility, non-oil generators called "Qualifying Facilities." Utilities were required to buy electricity from QFs that produced solar, wind, biomass and geothermal power. Most QFs, however, were small scale gas-fired cogeneration plants.

What does this have to do with the U.S. deregulation story? The QFs helped reduce oil-fired generation, but they also produced power at significantly higher average cost than the coal and hydroelectric generation already in place. Independence comes with a price. By the late 1980s, QFs had become more prevalent in the major industrial states, where demand had been growing faster. One third of New York's power, for example, was eventually supplied by high-cost QFs. California, New Jersey, Michigan, Massachusetts and Pennsylvania were also QF-heavy, pushing average power prices up. The QFs noticeably contributed to widening regional price differences.

The extra costs of the QFs, though, ended up being small compared to the costs of the dozens of nuclear plants being brought on-line in the wake of Nixon's 1973 exhortation. At one point, there were plans to build over 1,000 nuclear plants throughout the U.S. These plans were all shelved in March 1979, after the Three Mile Island reactor accident in Pennsylvania. No more nuclear plants were ordered in the U.S. after that, but much economic damage had already been done. Just as in Ontario, U.S. nuclear stations brought with them tremendous cost overruns and huge, unexpected debts that had to be incorporated into regulated rates. Utilities that chose not to invest in nuclear power were able to keep their power rates low. State-to-state power price differences became even more pronounced as regulators in industrial states, fearful of running out of power, pushed willing utilities to build even more new capacity, mostly coal-fired, than would be justified by rising demand. By the beginning of the 1990s, surplus capacity was estimated to be as high as 30 percent above peak demand, twice the traditional safety reserve margin of 15 percent. The cost of this excess surplus naturally went into power rates.

Large industrial corporations, such as automakers, steel mills, petrochemical plants and refineries, complained loudly about regional power price differentials. Lower-priced, mainly coal-burning states, such as Kentucky, Georgia and Texas, began to woo industry away from the higher-priced nuclear and QF states. But this was no help for factories already in place in higher-cost states. They began to ask some obvious questions: "Why should we have to move to Kentucky? Why can't we just buy our power from Kentucky, or from wherever it's cheaper? There's a transmission system in place. Let's just ship in the power. And what's happening with all this surplus power? In our businesses, when supply exceeds demand, prices come down, right? Why isn't this happening with electricity? What these overly protected utilities need is a little competition."

Thus began the U.S. electricity deregulation movement, the offspring of geopolitics, oil and the out-of-control nuclear industry.

At the beginning of the 1990s, deregulation was hardly a new idea in the U.S. Conservative economists and business schools had perennially advised governments to relax their hold on industry. Unleash the creative and entrepreneurial energy that now lies buried under a mountain of antiquated regulations, they said, then sit back and enjoy the show. Numerous benefits will inevitably follow: lower prices, technological innovation, higher productivity, better services. President Carter took the first giant leap of faith off this cliff when he deregulated the commercial airline industry, the railroads, oil prices and interest rates, as well as partially deregulating the interstate trucking industry. Carter also set up much of the deregulation machinery used by succeeding presidents Ronald Reagan (1981-89) and George H. W. Bush (1989-93) to sweep away or sharply reduce regulation in the savings and loan, banking, telecommunications, media, cable TV and other sectors. In those areas where regulation remained seemingly strong for cosmetic political reasons, such as in occupational health and safety and environmental protection, federal watchdogs were put on tight leashes and starvation diets—backdoor deregulation. Whether the impacts of all this deregulation were on balance positive or negative is still debated. Today, the majority view of consumer representative organizations in the U.S. seems to be that the downsides of deregulation—declining service levels, oligopolies, widening price and service disparities and a pronounced decline in genuine competition—far outweigh the alleged benefits of "getting government off the backs of the people."

One example of an unmitigated deregulation disaster that has relevance to what later happened in the electricity sector is Ronald Reagan's 1982 deregulation of the Savings and Loan (S&L) industry. After 1982, with virtually no government oversight and greatly increased federal deposit insurance, S&L managers went wild with depositors' money, investing in everything from highly speculative real estate ventures to kitty litter mines. Many of these investments were insider deals. S&L managers would set up companies, often with relatives as employees and directors, and then loan themselves the money. Organized crime moved in with plenty of investment "opportunities" of their own. When the investments failed, as many were intended to (so that the money would just "disappear"), U.S. taxpayers acquired the obligation to reimburse the fleeced depositors.

Over 3,000 S&Ls went under during the Reagan-Bush years, and over 500 commercial banks caught up in the S&L collapse failed as well. President Bush's son, Neil, brother of current president George W. Bush, helped run a Colorado S&L into the ground, costing U.S. taxpayers $1 billion. He never served a day in jail. The sentences given to those who *did* go to jail for S&L fraud were typically one-fifth that of the average bank robber. By 1986, a mere four years after S&L deregulation began, it was obvious that there were huge problems in the industry. The White House did not want to make an issue of it and possibly harm Bush's chances of being elected in 1988. Finally, after the election, Bush said he was shocked to learn of all the S&L perfidy, and moved to clean up the whole mess. The taxpayer bailout is being spread out over many years; estimates of its ultimate cost are as high as $1.4 *trillion* dollars, making it the largest theft in history.

I relate the S&L story for two reasons. First, it illustrates how effectively corporate interests in the U.S. were able to keep the idea of deregulation alive, even in the face of its most spectacular failure. Certainly after the S&L debacle the whole idea should have been discredited, or at least subjected to intense Congressional scrutiny. But instead of being presented as unimpeachable proof of the hazards of letting businesses do whatever they pleased, the S&L catastrophe was papered over by federal deposit insurance – most depositors got all their money back – and a long-term amortization of the huge public debt left behind. By 1992, the last full year of the senior Bush presidency, the real-world experience with deregulation was clear: call off the watchdogs and prepare to have the house looted. That was also the year that electricity deregulation began in the United States. It was an astounding triumph of ideology over experience, just like in Ontario. The second reason the S&L story is relevant is because a similar public bailout of the private utility sector became a feature of U.S. electricity deregulation.

In November 1992, as the last major act of his presidency, George Bush *père* signed the first comprehensive federal energy policy legislation since 1978. The Energy Policy Act, 1992, opened up the country's vast, 250,000 mile-long interconnected transmission network to all generators. This plan differed from the arrangement made under 1978's PURPA, which only required local utilities to open up their transmission lines to QFs selling power to that specific utility. Under the Bush approach, any generator, QF or not, could sell power to more remote purchasers and use the transmission lines of several utilities to deliver it. For example, an independent generator in Kentucky could use the transmission systems of the utilities of neighbouring West Virginia, and then Pennsylvania, to ship low cost power into New York. The Act also made such independent generators subject only to federal, not state, regulation.

Echoing presidents past, Bush praised the law as a way to reduce the nation's growing dependence on foreign oil. But insofar as it pertained to the electricity industry, the new law had nothing to do with kicking the U.S. addiction to imported oil and encouraging fuel diversity. The use of oil for electricity generation was already minimal. There was plenty of other generation available and there was plenty of cheap domestic fuel to ensure electricity security at reasonable prices for the indefinite future. Regrettably, that fuel was coal, but it *was* there, and in huge amounts. The United States is the Saudi Arabia of coal, with a supply that will last at least 250 more years. Bush was no tree-hugger, to put it charitably, so a concern for the environment was not his reason for opening up the grid to interstate wholesale competition, any more than it was to depress oil-fired generation. The real reason was to set the stage for a complete deregulation of the electricity industry throughout the United States, down to the smallest customer. This would wrest control of the electricity industry from 50 different state regulators and effectively dismantle the system of public accountability that had been in place since the New Deal. Even though state public utility commissions still had jurisdiction over domestic electricity transactions, it was believed that wholesale competition would inevitably lead retail customers to demand equal access to lower-priced power in other jurisdictions. It would only be a matter of time, deregulation advocates believed, before markets prevailed.

At first, the incumbent generating utilities were wary of deregulation, especially if they owned debt-ridden nuclear plants, as many did. What would happen to their companies if they were stuck with higher-priced, uncompetitive plants? These assets, and the debt that they carried, would be "stranded" in an open marketplace. Shareholders would lose many billions of dollars. A 1995 report by Moody's Investor Services found that half of the 114 utilities it examined had enough debt to put them at risk in an open marketplace. Surely, the utility lobbyists pleaded, it wasn't fair to punish their clients for building power plants that state regulators had approved. We're all in favour of competition, they said, but shareholders will have to be compensated if a political decision to introduce competition makes our companies uncompetitive.

Consumer groups and many politicians, however, smelled another S&L bailout. They suspected that either ratepayers or taxpayers would be stuck with the cost of the (mostly) nuclear mistakes. They asked: "Why shouldn't utilities write down these uncompetitive assets over time, like other failed businesses? Yes, this will depress shareholder profits for a few years, but why should ratepayers pick up the whole tab for the poor judgment and bad management of the past?" Of course if the money spent on nuclear plants had been invested in energy efficiency instead, perhaps there would have been no stranded debt.

I won't go into the distressing details of how the utility industry won the contest for getting ratepayers to pay the nuclear piper. With so many billions of dollars at stake, you can imagine how intense was the lobbying of state legislators, and how generous were the campaign contributions from the utilities. In the end, most state deregulation legislation provided for a high proportion of stranded debt recovery. Under California's restructuring law, for example, customers were charged a total of $10 billion over four years through a Competition Transition Charge on their bills. New York, Connecticut, New Jersey, Michigan, Illinois, Ohio, Pennsylvania, Texas and other states, passed similar enabling legislation. The same thing happened with Ontario's nuclear debt, as a prelude to the privatization of Ontario Power Generation planned by the Conservatives. Check the Debt Recovery Charge line on your hydro bill. This is Ontario's version of California's Competitive Transition Charge. Taking much of the old Ontario Hydro debt off the books of OPG would, the Conservatives hoped, make it easier to sell this public asset to private investors. A privatized OPG would not then be burdened with so much debt repayment and could therefore keep rates—and profits—that much higher.

While it isn't as large as the S&L bailout, stranded asset recovery in the U.S. will ultimately transfer many billions of dollars from ratepayers to utility shareholders. The U.S. consumer group, Public Citizen, estimates that the utility bailout could total over $200 billion. The public rationale for handing electricity customers the stranded debt was always the same: ratepayers would win in the end because the entire industry would then be more competitive, meaning lower prices and better services. "Don't hold your breath," warned many consumer advocates. This would turn out to be good advice.

14

The American Disaster—Part 2

Things have gone so badly for deregulation in the U.S. that less than half of all states have opened their retail markets and most of the others wish they never had.

When veterans of the U.S. Deregulation Forces hold their reunions in the future, they'll no doubt reminisce about the glory days of 1996 to 1997, when everything was going their way, when there seemed no doubt that they were winning. It was a brief but golden moment in time for the true believers of power deregulation. In the war of ideas, the complete victory of open markets over public control seemed at hand. Electricity deregulation bills clogged the hoppers of state legislatures; lobbyists clogged the halls. Polls showed most Americans believed competition would lead to lower prices and better services. Utility executives were lionized. "I told you so," academics told us so all over again. Power industry gatherings teemed with engineering, financial, operational, administrative, legal, political and marketing consultants, all looking for their own little sliver of a pie that seemed to be growing right in front of everyone's eyes with each chart-crammed Powerpoint presentation. The occasional sceptic who wondered aloud where the money was going to come from to pay for all these new industry players, was drowned out by deregulation cheerleaders. The sheer magic of the marketplace, went the cheer, would result in such remarkable efficiencies that consumer prices would fall and services would improve even after all the lawyers, the lobbyists, the bankers, the brokers, the aggregators, the analysts, the marketers, the traders and even the door-to-door salespeople were paid.

It was a heady, exhilarating time and the euphoria swelled with every legislative and regulatory advance. On April 24, 1996, the Federal

Energy Regulatory Commission issued its long-awaited Order No. 888, which prescribed precisely how the nation's entire interconnected transmission system must be open to all generators. Many in the industry marked the day with champagne. Completely open wholesale market competition was here at last, right there in black and white. An electrical Rubicon had been crossed. There was no going back.

In July of that same year, Colorado Congressman Dan Schaefer unveiled his "Electricity Consumers' Power to Choose Act," the first attempt at federal legislation to deregulate retail competition nationwide. It would require the states to open up their retail markets by December 2000 or have FERC do it for them. The bill also sought to repeal the Roosevelt-era Public Utility Holding Company Act (1935), clearing the way for the rebirth of the now all-but-forgotten utility "trusts" of Samuel Insull and J.P. Morgan's day. Indeed, the industry was already racing back to that time. Utility mergers and acquisitions activity took off like a rocket. Between 1995 and the end of 1997, over $300 billion in M&A transactions had either been completed or were in progress. The head of Hydro Quebec publicly predicted that the North American power industry would soon consist of only five or so mega-utilities. There was even speculation that Ontario Hydro itself, until recently the continent's largest utility, would be swallowed up by one of these giants. Ontario's Power Workers' Union ran newspaper ads promoting the idea of what it called "Hydro Canada," a merger of the Quebec, Ontario and Manitoba hydro systems that could dominate markets in the U.S. Northeast and Midwest. Manitoba Hydro scoffed at the idea. Hydro Quebec politely declined comment, saying it was a government issue.

In September 1996, California Republican governor Pete Wilson signed Assembly Bill 1890, which would give every customer of the state's three incumbent utilities the right to choose their electricity supplier as of January 1, 1998. The largest state, with its $23 billion a year power marketplace, wanted to be the first to fully embrace competition. True, the market design seemed complex and yes, consumer groups howled at the multi-billion dollar nuclear bailout, but AB1890 had passed the Assembly unanimously. Every lawmaker seemed convinced that it would bring power prices tumbling down and make California industry, which had been slipping in recent years, much more cost competitive. The bill gave an immediate 10 percent discount to small customers who stayed with their utility and froze rates until the stranded debt was retired or until April 2002, whichever came first. Over 200 marketers rushed to set up shop in the Golden State, apparent proof that there would be vigorous competition. The municipally owned Los

Angeles Department of Water and Power, however, had the right to opt out of the competition and chose to do so. That decision, controversial at the time, would end up saving Los Angeles consumers billions of dollars.

Other states, large and small, introduced deregulation laws during this time, with varying schedules for implementation. Maryland, Illinois, Arizona, Oklahoma, Maine, Michigan, Pennsylvania, New Hampshire, Ohio, Washington, New Mexico, Arkansas and others, jumped on the bandwagon. Some states wanted to take the California route and open up competition to all customers at the same time. Many, however, planned to phase in competition along the lines of the U.K. model. Large customers would be the first to get choice. If that worked out, small business and residential customers would follow. Almost every state provided for a mandated rate reduction or price freeze for residential customers, to allay voter fears that competition would actually increase their rates. A number of states that didn't pass deregulation laws during the 1996–97 glory years, were nonetheless moving in that direction. By 1998, 22 states, nearly half, had either opened their markets or had committed to do so. Collectively these states contained 60 percent of the population of the United States.

As January 1, 1998, approached, the Deregulation Forces seemed unstoppable. It mattered not at all that regulated power prices had, in fact, been dropping. At the end of 1997, before any state allowed customer choice, the U.S. Department of Energy reported that average retail electricity prices in the U.S. had declined for four years in a row. Reasons given for the steady drop included: lower costs of generation due to price drops in coal; an increase in hydroelectric generation over fossil fuel generation; lower interest rates and lower labour costs due to downsizing and mergers. Regulation had ensured that customers got at least some of the benefits of those decreased costs. Deregulation would soon put a stop to this.

Retail electricity supply competition in the U.S. was a bust from the moment it began. In California, less than one half of one percent of the state's 11.4 million residential customers changed suppliers. Most of the switchers chose Enron, which had invested $10 million in an aggressive pre-competition marketing campaign. In late April, only three weeks after market opening, Enron abruptly closed shop and left the California residential market, though it continued to sell to larger customers. Enron blamed the mandated 10 percent rate cut for its inability to crack the residential market and warned other states planning the same thing that "artificially low" rates would depress competition.

Critics noted that Enron, like everyone else, had known about the 10 percent reduction since 1996, but had chosen to uncritically believe industry analysts' predictions that customers would desert their traditional utilities in droves. Even the most pessimistic estimates said the switching rate would be around 20 percent in the first year; the optimists said it would be more like 50 percent. When quizzed about the poor response to customer choice, some California lawmakers sidestepped the issue by claiming that competition "is not just about lower prices," but also about innovative products and services, though they were hard-pressed to name any that were not already available.

On the other side of the country, Rhode Island and Massachusetts had also decreed across-the-board 10 percent rate cuts as part of their competition legislation which took effect in early 1998. Just like in California, customers back East greeted the exciting prospect of getting to choose their own energy supplier with a big yawn. As more state markets opened up throughout 1998-99, that yawn proved infectious, to the consternation of the cheerleading squad. No one was coming to the deregulation victory party. The champagne was getting flat. A hugely expensive effort to establish the first national utility marketing brand, *EnergyOne*, was put on hold until consumers started showing interest in competition. The idea eventually died without fanfare.

The first sign that the open market system was just a bit risky for consumers came in late June 1998, but not in California. Over the course of an entire week, Midwest spot market prices climbed as high as $7,500 per MWh, 250 times the pre-deregulation average price of about $30. At one point during the week, Chicago's Commonwealth Edison paid $4 million for $100,000 worth of power. Some trading company losses were so high they went out of business. Then, just as abruptly as they had gone up, Midwest prices settled down. The unprecedented price spikes were blamed on the weather. Mother Nature was apparently responsible for the high prices, not the deregulated marketers who demanded those prices. Though it was just as hot in southern Ontario, prices here did not go up as they were still regulated. Ontario Hydro made millions, however, selling Americans our surplus power.

Two weeks later, the puzzling price spike phenomenon hit California when spot market prices in the southern part of the state soared as high as $5,000/MWh, prompting the California pool operator to impose a price cap of a mere $500/MWh. "Congestion" on the grid, not hot weather, was blamed for the sudden jump. Consumer advocates pointed out that Californians had been paying pre-deregulation prices of $24 or less. Congestion hadn't been a problem back then. Why was it a problem now? And why wasn't it a problem every day? Power demand had been as high on many days before and after the spike. Something seemed fishy, but no one could explain the spikes. Nor could anyone explain why, a few weeks later, prices for reserve power in the state went as high as $9,999/MWh.

Alarmed by these early signs of market instability and still furious over the stranded debt charge, consumer groups in California joined forces to collect half a million signatures to put Proposition 9 on the November 1998 ballot. Proposition 9 would repeal most of AB1890, cancel the nuclear bailout and give all residential and small commercial customers a 20 percent rate reduction. A million dollars was spent on the "Vote Yes on 9" campaign. The utilities, in response, spent $30 million—less than one half of one percent of the $7 billion they had at stake—on a saturation ad campaign that portrayed a post-Proposition 9 California as much worse off. Don't be fooled by all this anti-deregulation propaganda, the utilities warned. Proposition 9 would inevitably lead to skyrocketing prices and even blackouts. Not wanting either of these, Californians defeated the ballot initiative by three to one. A similar referendum in Massachusetts also failed, for the same reasons.

In Washington D.C., enthusiasm for deregulation began to wane. The Schaeffer bill had been replaced by one proposed by Alaska Senator Frank Murkowski, but it too ended up going nowhere.

By late summer 1998, the bloom was already off the rose. The U.S. Energy Information Administration's weekly newsletter reported that "customers are not seeing the benefits of deregulation." The North American Electric Reliability Council (NERC), which had been set up after the 1965 U.S. Northeast blackout, reported a disturbing trend towards more power shortages. Several states that had been studying deregulation put the idea on the back burner. Others with legislation already passed, but not implemented, started talking about delaying the introduction of competition. In less than a year, the U.S. experiment with deregulation was wilting.

Don't worry, said the somewhat subdued, but still persistent, deregulation cheerleaders. This is a transitional period. We just need to do some fine-tuning of the market mechanisms. We're learning as we go. The rules need to be tweaked, but the idea is sound. The price spikes and shortages are sending the right price signals. Investors are eager to get into the game. This is only a temporary situation. Things will get better.

Instead they got worse.

In June 1999, rolling blackouts in New England left half a million people sweltering in the dark. In early July, power in upper Manhattan failed, driving people out of their oven-like apartments to sleep in the streets. One side effect: Columbia University medical researchers spent months reworking the millions of dollars in experiments that had been damaged by the outage. New Jersey, Pennsylvania, Delaware and Maryland were also hit. Later that month, to keep its entire system from shutting down, the utility that served Arkansas, Louisiana and southeast Texas, resorted to rolling blackouts affecting over 300,000 customers. To prevent even more blackouts, the utility paid as much as $2,500/MWh for energy that usually cost about $100 in the summer. Customers later got huge bills to cover these costs. In August, a series of blackouts plagued downtown Chicago, bringing billions of dollars of business in the Chicago Mercantile Exchange to a halt. Several heat-related deaths due to lack of air conditioning were blamed on the outages.

The extreme heat of that summer was the most obvious cause of the unusual rash of outages, but a number of industry observers said deregulation was part of the problem. Aging transmission and distribution infrastructures were not being adequately maintained or refurbished by utilities now more focused on "competitiveness." Wires were a boring business, power trading was where the action was. Moreover, utilities with transmission lines were increasingly reluctant to divulge important operational information that they had shared in the past with neighbouring transmission owners. In a deregulated marketplace, such information could have commercial value. In testimony before Congress in November 1999, David Nevius, vice president of the North American Electric Reliability Council, warned that under the newly competitive system, "We may not be able much longer to keep the interstate electricity grids operating reliably." Under the old, regulated system, he said, transmission system users and operators cooperated voluntarily to ensure reliability. "Now, however, they are competitors and don't have the same incentives to cooperate." Nevius could have added that more than 150,000 U.S. utility workers had lost their jobs in the wake of deregulation. Power systems are not self-maintaining. Something's got to give if you neglect their upkeep. Something did.

In the heat of the summer of '99, many who had once supported deregulation began to get cold feet. The Nevada legislature had voted in 1997 to open up its market by the end of 1999. But in June, the same lawmakers voted to delay competition and gave the governor the authority to decide when it would begin. The deadline for competition was extended several times; eventually the idea was dropped altogether. In April 2001, Nevada became the first state to officially reverse, not just put on hold, its deregulation plan. Governor Kerry Guinn also put a moratorium on the sale of the state's power plants.

Nearby New Mexico had passed a deregulation law in 1999, but cautiously delayed its implementation because of growing concerns that it wasn't working elsewhere. In March 2001, the state delayed competition until 2007 and stopped the planned breakup of its main utility, Public Service of New Mexico.

In November 1999, a Colorado panel recommended against deregulation because a study had predicted that the state's already low power prices would go up substantially. The reason was simple to understand: Colorado generators would be free to sell their output into the wider open market where prices were higher, often much higher. Coloradans would then be forced to compete for their own power against customers in nearby states, including California, where prices were more than twice as high. You didn't have to be a rocket scientist to figure out that if you give your generators the opportunity to charge more for their power, they'll do it. Which takes us to the sad story of Montana, due north of sensible Colorado.

Montana, with its abundant hydroelectric resources and mountains of coal, used to enjoy regulated power prices that were the sixth-lowest in the country, almost as low as Colorado's. But in 1997, large Montana industries lobbied hard for deregulation, spurred on by the privately owned Montana Power Company, which had its eyes on those lucrative California markets. "Don't worry about supply," Montana Power told its large industrial customers. "There's plenty to go around. Your prices will come down—guaranteed." A lot of money flowed into the Capitol building in Helena and those who had been plumping for deregulation got their wish. Wholesale prices were freed up in 1998, retail choice would come later. For the first year or so, wholesale prices stayed relatively low. Except for the occasional price spike, California markets had not yet heated up. Then, in December 1999, Montana Power sold all its generating stations, including several dams, to Pennsylvania-based PPL. It kept the regulated wires businesses for itself. Within weeks, shortages mysteriously developed and wholesale electricity prices in the state went up as much as 400 percent above the previously regulated rate. Wild swings in wholesale prices mesmerized and frightened Montanans. Why was this happening? What was going on?

Defenceless Mother Nature was blamed again, this time for cold weather, but people were having none of that. It had been just as cold last winter and there was no change in generation capacity in the state. Politicians ran for cover. PPL's headquarters back East issued bland assurances that the problem was "temporary." It wasn't. Sky-high prices persisted. Many industries that had lobbied for deregulation were forced to shut down plants, mines and smelters. Well over a thousand primary industrial workers lost their jobs. At least an equal number of secondary, service sector jobs were lost as well, probably more. That was a lot of jobs in this sparsely populated state. Montana's 147,000 square miles contain only about one-third the number of people who reside in Toronto. "We've had some bumps and grinds," the head of Montana Power admitted in late 2001. But, he said "it's going to settle out, ultimately, as more supply is built." Since the state already had far more generating capacity than it needed for its own use, it was unclear how even more would help.

Montana's experience provides a very important lesson to Ontario. We're told by the Conservatives that deregulated wholesale rates are rising here because there is a shortage of power (due to our unreliable nuclear stations which previous Conservative and Liberal governments built). When we demand more power than we can generate, prices naturally go up. "It's supply and demand, stupid, what did you expect?" is the Conservatives' pat response to questions about our power price jumps here. "Just hang in there until we get more supply and then you'll see the benefits of deregulation." The case of Montana shows the folly of this ideological cant. Demand in Montana is only 40 percent of what its domestic generators can supply, a phenomenal surplus. According to free market theory, their prices should have come down once competition was introduced. The very opposite has happened, in spades. And we know exactly why this has happened. It's supply and demand, stupid. As Colorado figured out, once Montana freed its generators from the obligation to serve its own residents at regulated rates, the state's entire power production capacity became available to tens of millions of customers in nearby states whose collective demand for power greatly exceeds Montana's supply.

The very same thing will happen to Ontario, which for nearly a century enjoyed lower power rates than most U.S. states, including New York and Michigan, our biggest export customers. Most of that advantage comes from our extremely low-cost hydroelectric power. Thanks to the Conservatives' blind faith in deregulation, however, Ontario consumers no longer have the right to sole benefit of that hydropower. It must be shared with the Americans, who now have the right to outbid us for our own power. The power from Niagara's Sir Adam Beck stations costs, in U.S. currency, less than half a cent per kilowatt hour to produce. That's what Ontarians used to pay for it. If hydro privatization and deregulation continues, we will have to pay what the

Americans are willing to pay for it, which is one of the reasons our power prices have gone up so much since deregulation. God bless the Americans, but they should not be allowed to strip away, every single day, the wonderful benefits of a great public power system, the work of five generations. It is painful to think of Adam Beck looking down on us now, saddened or outraged, probably both, to see that the great hydroelectric stations that bear his name will be used to reduce American power costs at the expense of Ontario. Now *that's* something worth voting against.

In May 2000, California wholesale prices suddenly jumped to more than $500/MWh. Customers in the San Diego region, in the very southern part of the state, were stunned by power bills in June and July that were double, then triple what they had been in the spring. Their utility, San Diego Gas & Electric (SDGE), had worked its way out from under the legislated rate freeze by selling off some of its assets for good prices and using the money to pay off its stranded debt bonds early. SDGE could now legally pass on the full cost of wholesale power price increases to its customers. Customers of the two other state utilities, Pacific Gas & Electric (PGE) and Southern California Edison (SCE), were still protected by the freeze. PGE and SCE could only charge their customers around six cents per kwh ($60/MWh) for now deregulated power that cost the utilities eight times as much.

San Diego power bills became national news; deregulation was getting some very bad press. To calm the situation, the California Independent System Operator slapped $250/MWh price caps on wholesale power sold in the state. In retaliation, generators then began selling more power outside of California, where no caps existed. Rolling blackouts were imposed in the northern part of the state. The transmission system operator declared Stage 3 alerts almost daily, meaning that electricity reserves had fallen below 1.5 percent. By the fall, wholesale prices in the western U.S. reached as high as $1,400/MWh. The market was spinning out of control. Retail prices in San Diego were re-regulated, retroactive to the previous June. The state's utility deficits mounted by the hour, by the minute, in fact. In December, the Federal Energy Regulatory Commission stepped in and removed the state's price caps as a way to bribe California generators to sell their power to Californians. Prices moderated somewhat that winter, falling to an average of $300/MWh, but the utilities were still losing 24 cents on every kilowatt hour they sold to their customers. In April 2001, PGE declared bankruptcy; SCE was coming close to the same fate. Meanwhile, independent generators were carting trainloads of money out of the state. Reliant Energy, for example, registered a 600 percent increase in earnings during the third quarter of 2000, courtesy of California.

The situation continued to deteriorate as power supplies evaporated into thin air, even as prices rose. Many manufacturers shut down their plants and sent workers home. Others shifted their production from day to night, in response to the market's new "price signals." Air emission limits were lifted as every diesel generator in the state was cranked up. California's once mighty power utilities were now credit risks who had difficulty buying power from independent or out of state producers. In January, Governor Gray Davis declared a state of emergency and authorized the state's Department of Water Resources (DWR) to buy power on behalf of the utilities. In effect, this "nationalized" power purchasing because it made California taxpayers responsible for the cost of the state's electricity. This did not relieve power shortages, however. By mid-year 2001, the system operator had declared 38 Stage 3 Alerts. Most of these emergencies were accompanied by rolling blackouts of 60-90 minutes each so that electricity could be rationed around the state. It is important to remember that prior to June 2000, power emergencies in California were rare and blackouts from power shortages had been non-existent.

California's woes severely affected its neighbours, especially the Pacific northwest states of Oregon and Washington, from which California had long drawn cheap hydroelectric power. Lower rain and snow levels in 2000 and 2001 led to declines in these states' electricity resources. Competition for the power from California "naturally" drove up regional wholesale prices. In early 2001, Seattle City Light raised rates by nearly 30 percent. The federally owned Bonneville Power Administration said it would have to raise wholesale rates as much as 250 percent by the fall unless things turned around.

In an attempt to alleviate California's power shortages and prevent industry from deserting the state, the Department of Water Resources signed $43 billion worth of long-term power supply contracts averaging about $70/MWh, about twice the pre-deregulation price. To pay for these contracts, the California Public Utilities Commission (CPUC) ended the electricity price freeze, raised rates 40 percent, and took away the right of customers to choose their power supplier. Deregulation was dead in California and was dying in the rest of the West. In December, the Federal Energy Regulatory Commission imposed a still very high wholesale price cap of $92 on the entire 11-state Western U.S. region, an admission that the open market was a failure.

The total cost to California of its adventure with deregulation, when lost industrial production and other economic impacts are added up, has been calculated at $45 billion. A bitter irony: during 2002, wholesale prices in California stabilized at near pre-deregulation levels, leaving the state government holding the unattractive-looking bag of high-priced contracts it had signed to protect its citizens. On occasion, the Department of Water Resources had to sell for $1/MWh power it had bought for $70.

During the California energy crisis, and for months afterward, explanations of its causes became an industry in itself. Deregulation disciples retroactively spotted all kinds of flaws in the market design. Academics churned out papers filled with Greek letter-laden equations purporting to pinpoint the problem. Private power industry associations variously blamed rising natural gas prices (much of California's generation is gas-fired) and inadequate transmission facilities. Republicans blamed Democratic Governor Davis for fumbling the ball; Democrats blamed former Governor Wilson for signing the legislation in the first place. Conservative commentators blamed liberal "tree-huggers" for resisting the construction of more power plants and transmission lines. The mystery that none of these explanations could solve, however, is why, after competition was ended, had prices dropped precipitously and blackouts stopped?

The answer came in September 2002, after an exhaustive investigation by the California Public Utilities Commission. During the shortages, the CPUC said, the state's five largest power marketers deliberately generated an average of 40 percent less energy than they could have. The companies named were Duke, Mirant, Dynegy, Reliant and Williams/AES, all five of which have a presence in Ontario. According to Gary Cohen, CPUC attorney and one of the report's authors: "(The companies) knew the system was in an emergency and they thought that by withholding capacity, they could drive prices up. It wasn't a question of California not having enough generating capacity. That's the story we heard. This report shows that those excuses were untrue." Naturally, the power companies denied the accusations, but had a difficult time explaining away their own production and outage records. Curiously, the power shortages and high prices suddenly ended when the government began to subpoena those records.

This is not to say that California does not need more generation capacity or much more energy efficiency; it needs both. Nor was the withholding of power the only explanation for high wholesale prices. The market was manipulated in other ways, such as the now-famous "round trip trades" developed by Enron and others. In these bogus transactions, power produced in one state was "sold" to an entity in a neighbouring state, which then sold it back to the state where it originated. As I write this, Ontario's Independent Electricity Market Operator is investigating the possibility that this same tactic was used to push up power prices here during our first summer of deregulation.

Before the West Coast meltdown, the Ontario Conservative government frequently cited California as a model of the consumer benefits of power industry competition. When it finally dawned on the Conservatives that the California experience was validating the NDP's opposition to deregulation, they abruptly changed their tune and began singing the praises of deregulation in Pennsylvania, where residential prices were lower than they had been under regulation. "That stuff we said about California? We really meant to say Pennsylvania. Must have been a typographical error. Besides," they went on, "Ontario is not at all like California. For one thing, California has a shortage of generation, which is why things went crazy there. Here in Ontario, we've got plenty of power. What happened in California couldn't possibly happen here. We're like Pennsylvania, which also has a generating surplus. Rates are coming down there. See? It *works*."

The Pennsylvania story, though, turns out to be yet another example of why deregulation doesn't work, at least not for consumers. Pennsylvania began opening up its retail marketplace in 1999 and, as in California, dozens of hopeful marketers flocked to the state to flog their supposedly better, or at least cheaper, electrons. As in California, interest was low at first, but within a year over seven percent of the state's customers had signed up with an alternative supplier. In return for getting their stranded debt taken care of, the two incumbent privately owned utilities, GPU in the north and Peco Energy in the south, were made the "default" providers for small customers who didn't want to shop around. GPU and Peco were obliged to provide these default customers with power at 4.528 cents per kilowatt hour (6.8 cents Cdn.), which was 10 percent below the previously regulated price. Default price caps in Pennsylvania will be in place for several more years.

"All hail the Pennsylvania success!" the Conservatives fairly shouted in the Legislature in 2001. "Consumers are winning. Competition works!" They never mentioned that the rate reductions in Pennsylvania were legislated, that deregulation was "working" because of regulation. Nor did they reveal that pretty well all of that 10 percent reduction was eaten up by the nuclear stranded-cost bailout which was being financed by the Competitive Transition Charge and something called the Intangible Transition Charge. Both charges were added to customers' power bills. Nor did they talk about the fact that Pennsylvanians were still paying 50 percent more for their power than we were here, in regulated Ontario.

But even as the Conservatives were animatedly pointing to Pennsylvania as fresh evidence for the wisdom of deregulating, the state's power marketplace was unravelling. A July 2001 report from the Office of Consumer Advocate revealed that the state's Electric Choice program was in complete disarray. Wholesale power prices on the Pennsylvania-New Jersey-Maryland (PJM) system had been rising for months and had doubled to nearly $120/MWh by late spring. At times, PJM wholesale prices reached $1000/MWh. The reason they didn't go any higher is because this was the FERC-allowed price cap. The new retailers, who had to pay these higher wholesale prices, were dropping like flies, unable to beat the default price. Their customers rushed back into the regulated arms of GPU and PECO. Where there had once been 30 competitors in GPU's service territory, there were now three. About that many remained in Peco territory. "Much remains to be done," said the state's official Consumer Advocate, Irwin Popowsky, towards developing a healthy, competitive market in Pennsylvania.

The story doesn't end there, however. GPU and Peco were also losing money fast as the spread between the unregulated wholesale and regulated default retail prices widened and persisted. In 2001 alone, both utilities lost hundreds of millions of dollars protecting consumers from the benefits of competition. But that shield is only temporary. The Pennsylvania Public Utility Commission is allowing a surcharge on power bills to cover the losses. Once again, as with the transition charges, the surcharge will not be reported as part of the energy rate, preserving for simple-minded Conservatives in Ontario the notion that Pennsylvania's open market is actually benefiting consumers. That surcharge in likely to grow by quite a bit in the years to come, until the price caps come off.

Shall I go on? Should I tell you the story of how Massachusetts, which also mandated a 10 percent rate cut for "Standard Offer" customers to make competition an instant success, had to actually raise price caps in order to attract competition to the state? How bizarre is *that*? Apparently, customers weren't going to benefit from competition unless they were willing to pay more. Should I tell you about the double-digit price increases in post-deregulation New York and how the state's Power Authority had to ask FERC for wholesale price-capping powers? Or about the ongoing lack of consumer interest in "the right to choose" in Ohio, Iowa, New Jersey, Rhode Island, Michigan, Texas and pretty well everywhere else where deregulation is in effect? Or about the fact that half of all U.S. states have firmly decided not to proceed at all with deregulation and that several who have already gone down that path are now in various stages of retreat? You already know about the 2001 collapse of Enron. You may not know that scores of other U.S. power marketers have either gone under or withdrawn from trading since Enron's demise. Many are still being investigated by the Securities and Exchange Commission under the SEC's "It's about time we locked the barn door" program. You should also know that U.S. senators and governors from still-regulated states have been aggressively resisting the FERC's current plan to impose a uniform market structure on all lower 48 states through the creation of Regional Transmission Organizations. They want no part of this epic failure.

You won't be surprised to learn that most Americans consider deregulation a colossal failure. The Consumer Federation of America, an umbrella body of 285 consumer organizations representing over 50 million Americans, continuously implores governments to stop the madness, calling deregulation "all pain, no gain." While it's true that average power prices for large customers have declined since 1998, they had already been declining for four years before that, before competition. And, as we have seen, small consumers in deregulated states have only benefited because state-decreed rate cuts and smoke-and-mirrors transition charges mask the true costs of competition. In May 2002, for example, Illinois froze its residential power rates until 2007.

And where are all the eager entrepreneurs who were supposed to be tripping over each other in the mad rush to build more power plants to better serve Americans? Out looking for better investments, that's where. Share prices of U.S. electrical utilities and power marketers, once the darlings of Wall Street, are today a fraction of what they were two years ago, just like in the U.K. Private power credit ratings are at historic lows. It's tough, very tough, to get financing for new power projects. Plans for tens of thousands of megawatts of new capacity have been left on the shelf to gather dust since 2001. Even before September 11 and the Enron debacle, private power had been in retreat. The growing energy needs of the nation are being met by cranking up old, operationally cheaper and dirtier coal plants, just like in the U.K. The bubble of optimism that new, more efficient and cleaner plants would spring up has been burst, just like in Ontario.

The fact that deregulation in the U.S. is not working is silently acknowledged by the Conservative government of Ernie Eves. It long ago stopped talking about how consumers in this or that state are benefiting from competition. They know that there are no real examples to cite; all the old ones have been discredited. They've also stopped mentioning Australia and New Zealand, whose own experiments with privatization and deregulation have failed, unless power shortages, higher greenhouse gas emissions and higher prices are considered a success.

Such is the great power of Conservative ideology, however, that even when hit by wave after wave of evidence that disproves their blind faith, they cling to it with even greater tenacity, furiously inventing reasons—often on the fly, without any real analysis—for why the theory doesn't seem to be working out in practice. This, of course, is the same ideology that drives the Conservatives' approach to health care, education, environmental protection, corrections, highways, transit and other public services. Even when presented with evidence that their policies in these areas are not working as advertised, they deny that there is anything wrong with their thinking. "Just give us a little more time," they say, "and we will hammer these pesky realities into shape and make them all conform to our beliefs."

Well, they've had more than a little time. It's been eight years.

15

The Ontario Disaster—Part 1

The Harris Conservatives inherited an invigorated and financially healthier Ontario Hydro, but continued to insist that regulated public power was a bad idea.

On June 8, 1995 the Conservatives regained the government they had lost ten years earlier. Their timing couldn't have been better. Whereas the New Democrats in 1990 had to grapple with a sick economy, by 1995 Ontario's economy was roaring out of recession. In 1994, the last full year of the Rae government, 194,000 jobs were created in Ontario, 22 times the population of my hometown, Fort Frances. Optimism was on the rise. Business was once again flocking to the province and those already here were expanding. As a result, tax revenues were way up, and climbing; social assistance rolls were declining. The provincial budget deficit was shrinking fast. There was no pressing need for Conservative Premier Mike Harris to slash social assistance rates by over 20 percent, victimizing tens of thousands of children, among many other helpless people. There was no need to gut health care and education funding. There was no need to stop assisting urban public transit. There was no economic imperative to cut back on water inspections and enforcement of environmental laws. And there was certainly no need to abandon Ontario's publicly owned electricity system, which had turned around financially by 1995 and was bringing in hundreds of millions of dollars in profits. Those profits belonged to the people of Ontario, as they had since the days of Adam Beck.

The greatly improved health of Ontario Hydro was detailed by the soon-to-depart Chairman, Maurice Strong, in a widely reported speech he gave two weeks after the Conservatives took office. After reciting the dreary statistics he had inherited when he took over in late 1992, Strong said:

In early 1993, we took decisive actions to overcome this crisis. First, massive cost reductions to halt spiralling rate increases and to reduce Hydro's indebtedness. Second, a fundamental restructuring of Hydro's organization to make it more business-like, more flexible and responsive to customer needs. We made deep and extensive cost reductions. We committed to a no real rate increase ceiling for the balance of the decade, reduced staff by more than 30 percent and OM&A costs by $600 million and slashed capital spending to $13 billion over 10 years…we expect to be able to reduce our $35 billion debt by about $8 billion over the next five years.

Not exactly the basket case the Conservatives had been portraying before the election. Indeed, by the end of the year, Hydro's long-term debt stood at $31.4 billion. Strong's successor, William A. Farlinger, confirmed this assessment of a recovering Hydro a few months later. Farlinger, a close friend of Harris's and a top Conservative fundraiser, was appointed Hydro's new Chairman in November 1995. He had been the head of Ernst & Young Canada, Ontario Hydro's external auditor. He had also been a member of Hydro's Financial Restructuring Group in the mid-1990s. Farlinger knew perhaps better than anyone, certainly better than Premier Harris, what the true state of Ontario Hydro was and he was clearly no fan of the NDP. But as Hydro's new Chairman, he signed the 1995 Annual Report that declared the previous year, 1994, a record year for Ontario Hydro revenues; 1995 was just as good. The debt was coming down even faster than projected. Farlinger was, he wrote, "very optimistic" about Hydro's future. Perhaps he told the Premier something different while they were out on the golf course, but his public stance was that Hydro was in pretty good financial shape.

Farlinger, however, was also an open advocate of deregulation and privatization. He had co-authored a report released in August 1995 that warned of Ontario's dim economic future if it did not follow the Americans down the road to open markets for private power. Even though the report admitted that "Ontario Hydro's rates are currently lower than its immediately neighbouring U.S. states," all that would change when the U.S. fully embraced competition. Our industries would either leave us or start generating their own power, leaving Hydro in the lurch with huge stranded assets. Let's hitch our wagon to the U.S. marketplace, the report enthused, noting more than once that Ontario Hydro has "one of the largest surpluses in North America and low incremental costs from its hydro and nuclear units." We can sell gigawatts galore to the power-hungry Americans, Farlinger insisted, but *only if we privatize our system*. Why was that? Hydro's ability to capitalize on U.S. export opportunities, the report said, is "restricted by its status as a crown corporation." If we try to sneak more lower-cost, publicly owned power across the border, it would "open the threat of countervail and other retaliatory measures by U.S. producers, particularly against a government-owned organization."

It's not clear how Farlinger arrived at the conclusion that power from a publicly owned utility like Ontario Hydro would be unwelcome in the U.S. That has certainly not been the experience of publicly owned utilities like Hydro Quebec, Manitoba Hydro and BC Hydro, which have collectively sold billions of dollars worth of surplus power to Americans since the U.S. wholesale market opened in 1996. The U.S. Federal Energy Regulatory Commission (FERC) would bar us from open access to its country's interconnected transmission system if we didn't fully open up our own wires to U.S. generators, which is fair enough. But in the land of the Tennessee Valley Authority, the Bonneville Power Administration, the New York Power Authority and many other publicly owned utilities, it seems very unlikely that Ontario Hydro would be sanctioned as a "cheater" simply because it is government controlled. In fact, Ontario Power Generation was granted a FERC license as of May 1, 2002, the day the Ontario wholesale market opened. Farlinger was wrong, as was his linchpin argument for Hydro's privatization.

Hydro's operational and financial comeback was a mixed blessing for Harris and Ernie Eves, his friend and then Finance Minister. There were two obvious pluses. First, they could give some of the new money away and promote their hard-nosed, cost-cutting image by extending and enlarging the NDP's rate freeze. Secondly, a leaner, financially healthier Hydro would be more attractive to the private sector and therefore more likely to be sold for a politically acceptable price. The Conservatives were smart enough not to immediately engage in a Thatcher-style public asset giveaway, which ultimately proved the undoing of the U.K. Conservatives. On the negative side, the much improved health of Hydro also made it a more attractive public asset, at least the non-nuclear parts. Harris and Eves could already hear the objections to Hydro's privatization: "Why sell it, when it seems to be improving? Maybe Hydro went through a bad patch because of all those nuclear cost overruns. But look: the debt is going down, profitability is going up. We own those profits. Why sell Hydro when its value is rising?"

Indeed, that would have been a good argument even if everyone in the province had agreed that privatization was the way to go. No one was sure how the U.S. regulatory story was going to play out. There were a lot of potential stranded assets out there. What would they be worth? No one would really know the market value of Ontario's power system assets until the details of each jurisdiction's legislation were in place. Some real life experience with the open market was also needed. Guessing the market value of the Bruce Nuclear Station, for example, would have been a shot in the fog until Ontario's legislation was enacted and the market rules written. Even after that, it would still be risky. Farlinger likely advised Premier Harris to go slow on deregulation and privatization. "Wait until we get a better handle on what's going on in the U.S.," was the sensible advice, wherever it came from.

Nevertheless, Harris and Eves still had to be seen to be doing *something*. As pre-election privatization firebrands, they had encouraged the expectation that very dramatic change would quickly be visited on Hydro, either through privatization or competition or both. How to stall, without losing momentum? Fortunately, the sheer magnitude of the change, arguably the biggest political decision in Ontario history since 1906, warranted the appointment of an "independent" commission to study the matter. The government could then take its time studying the work of the studiers. Legislation could be delayed until the most politically propitious time. In late November 1995, Conservative Energy Minister Brenda Elliott announced the appointment of an Advisory Committee on Competition in Ontario's Electricity System. Its terms of reference were to investigate and "consult broadly on" a wide range of complex technical, financial, structural and regulatory issues and make recommendations on how to reshape Ontario's 90-year-old power system for the twenty-first century. Oh, and could we have that report by April 30?

Chairing the committee and lending it his name was high-profile Liberal Donald S. Macdonald. Macdonald, now in private practice as a corporate lawyer, had held several federal Cabinet posts during the Trudeau years, including Minister of Energy, Mines and Resources. He must have done a good job in that portfolio because after leaving government he showed up on the boards of several corporations with rather large interests in energy: TransCanada PipeLines, Alberta Energy Company Ltd. and Siemens Electric among them. When asked by a reporter if his corporate ties to the energy sector might be seen as a conflict of interest, Macdonald snapped: "If it's a problem, I'll resign."

The composition of the rest of the committee made the general outcome of its work a foregone conclusion. Darcy McKeough was a Minister of Energy in the Conservative government of Bill Davis and later head of Union Gas Ltd. Robert Gillespie was Chairman and CEO of General Electric Canada. Jan Carr was from Acres International, a power sector engineering consulting firm that naturally had an interest in anything that led to more electrical system construction activity. John Grant was former chief economist of Wood Gundy, the Bay Street investment firm. Leonard Waverman, a University of Toronto economics professor, was a big fan of deregulation and had written extensively on competition policy in the utilities sector. Joining the guys was Sylvia Sutherland, a former mayor of Peterborough with excellent mainstream environmental credentials. Besides environmentalists, was Sutherland realistically supposed to speak for the province's residential and small business consumers, municipal hydro commissions, farmers, First Nations and other stakeholders who weren't represented on the committee? The privatization and deregulation steamroller was on the road. Nothing was going to stop it.

Then a remarkable thing happened. Simultaneous with the announcement of the Macdonald Committee, the Power Workers' Union launched the largest, most expensive public campaign ever waged by a union in Canadian history. In late November, full page ads in daily newspapers across the province shouted "Watch Out Ontario, You're About to be Taken." The ad explained, in unusually rich detail for advocacy advertising, the financial turnaround at Ontario Hydro and the consequences of private power in the U.S. and the U.K. It accused Bay Street and Hydro management of pushing for "the biggest sell-out in Ontario history" and openly implied that Hydro managers were looking to get rich on huge pay increases and stock options, just like their U.K. cousins had done in that country's privatization. It was the bottom line of the ad, however, that best summed up the issue: "Control of Our Energy is Control of Our Future." The PWU had hit the nail squarely on the head.

That initial round of newspaper ads was followed by several others over the course of the next six months. Some of them further exposed the realities of privatization and deregulation. Others were history lessons, reminding readers that the people of Ontario had voted for public power by popular referendum and should have the right to do so again. "No Vote, No Sale," challenged one such ad, featuring a ballot meant to evoke the historic votes of 1907 and 1908. Another carried a portrait of Adam Beck and his famous deathbed quote about "forging a band of iron around the Hydro" to protect it from the politicians. The PWU also ran radio ads during rush hours and even television spots during the 1996 Stanley Cup playoffs. A PWU brochure explaining the hazards of privatization and deregulation was delivered to every household in every city visited by the Macdonald Committee. PWU-distributed bumper stickers admonished Harris directly: "Don't sell it, Mike. It's ours." Niagara Falls was made the central symbol of the campaign. Would the government be so base as to sell Niagara Falls?

It was a very effective campaign that generated much public attention, making Hydro a front burner issue for months. By the time the Macdonald Committee finally reported in June, only a few weeks behind schedule, polling showed public opposition to Hydro privatization had mushroomed in the past year, from 52 to 74 percent. The PWU campaign had worked. But would it matter to Harris? He had time on his side and was in no hurry to move on the issue. He did, however, stop talking about Hydro privatization for at least two years. His new Energy Minister, Norm Sterling, frequently denied that the government was considering a sell-off.

Macdonald's report, *A Framework for Competition*, called for a sweeping restructuring of the province's electricity system. Its main recommendations:

- Open up the marketplace, first for wholesale, then for retail customers "as soon as practicably possible" and end Ontario Hydro's monopoly.
- Split up Ontario Hydro into separate generation and transmission businesses.
- Privatize all the fossil and most of the hydroelectric stations. The Niagara stations, however, would be kept in public hands.
- Maintain public ownership of the nuclear stations (presumably because no one would buy them anyway).
- Give the Ontario Energy Board full authority to regulate the province's electricity industry.
- Consolidate the province's distribution system into fewer utilities and let the Ontario Hydro Retail distribution assets be absorbed by them.

All told, there were more than 50 recommendations. Only one mentioned the environment, blandly calling for "consideration" of environmental protection during restructuring. Sylvia Sutherland must have felt more than a little disappointed at the pointed lack of interest the committee had shown in the subject.

Within days of its public release, the Macdonald Report fell off the radar screen. Whether it was the effect of the PWU campaign, the still-uncertain situation with deregulation in the U.S. or the number of other fronts on which the Harris revolution was being fought, no one in the government would speak of the report beyond acknowledging its existence. Ontario Hydro itself would not comment. Industry publications analyzed it to death for a couple of months, but the mainstream media quickly lost interest. So did the public, who went back to taking Hydro for granted. It was no longer a front burner issue. Macdonald was unceremoniously shelved.

In the background, however, much was happening. Hydro was furiously reorganizing into its three "signature" businesses: generation, transmission and retail, each one promising to be even more commercially oriented than the others. The financiers began jockeying for position, in anticipation of the IPOs to come. Bay Street law firms scrambled to put together energy practice teams. American power companies quietly opened small offices in Toronto and hired Canadian, mainly ex-Hydro, representatives who dutifully networked at the endless string of energy industry conferences. Senior bureaucrats and political staff at the Ministry of Energy beavered away at briefing notes, policy papers and legislative proposals, always with one eye on their post-government prospects; surely all these new industry players would eventually need their unique expertise, insights and contacts? The government sent out feelers to the Power Workers, its most effective opposition on this issue. What would it take to get the union onside? Was this even possible, given Harris's open contempt of the labour movement and his apparent determination to displace Mitch Hepburn as the most anti-union premier in Ontario history?

In August 1996, the nuclear sins of past Conservative (and Liberal) governments once again strode onto centre stage and asked for everybody's attention. Ontario Hydro's nuclear division (OHN) was near meltdown. The constant warnings of the federal regulator, the Atomic Energy Control Board (AECB) that it might have to shut down some reactors unless operations improved considerably, had had little effect on OHN. Prescribed repairs were commonly months behind schedule. New surprises were always turning up. Labour-management relations in the nuclear plants, never particularly good, spiralled downward. Soft-spoken Hydro President Allan Kupsis, himself from the nuclear side of the business, met often with the forceful PWU President John Murphy, who had been a nuclear technician at Pickering and now enjoyed the unqualified support of his 18,000 members. What could be done to turn OHN around? Murphy told Kupsis what he already knew:

current nuclear management, from the head office to the station floor, was dysfunctional, incompetent and incapable of gaining the respect of the workforce. If you want change, change management.

Kupsis threatened to do just that if things didn't turn around, and fast. He publicly mused about shutting down the nuclear stations altogether, calling them "uncompetitive." Breaking with Hydro's tradition of disdain for the American nuclear industry, Kupsis brought in Gregory Kane, described as a "leading U.S. nuclear expert." Over $400 million was budgeted to fix the problem-plagued stations, which were then operating at only 70 percent of capacity. Kane didn't last long. In December, Kupsis brought in G. Carl Andognini, another "leading American nuclear power expert" who had worked on the rehabilitation of Three Mile Island. Taking charge right away, Andognini imported a team of other Americans to swoop down on OHN to figure out exactly what was wrong and how it might be fixed. Their report, delivered to the Hydro Board in mid-August 1997 was alarming. Nearly every paragraph was damning. Here's one, typical of the whole report:

> Each station had substandard equipment. Important equipment was out of service. Insufficient or improper maintenance resulted in leaks from mechanical joints. Operators, engineers and station management accept low standards for equipment performance and condition.

Almost everything about nuclear operations, the Andognini report said, was "Minimally Acceptable," one small step above unsafe. The team recommended shutting down the older Pickering A and Bruce A reactors in order to concentrate resources on getting the newer B stations and Darlington up to par. "Give us three years and $5 to $8 billion," said the Americans, "and we'll fix 'er right up." Kupsis took responsibility for the mess, fell on his sword and resigned from his $500,000 a year job. Chairman Farlinger publicly theorized that the nuclear division was being operated as a "cult" and that senior management had not been digging sufficiently into what was going on. Farlinger's apparent surprise at Andognini's revelations was understandable. He had only been chairman for 21 months. The board approved Andognini's plan to shut down 40 percent of Hydro's on-line nuclear capacity. The lost power would be made up by running Ontario's fossil plants full bore and by importing power from the U.S., pretty well all of it fossil, all of it expensive.

In the Legislature, an all-party Select Committee on Ontario Hydro Nuclear Affairs was quickly set up to do its own investigation of this latest nuclear mess. The committee heard from nearly 100 witnesses, many of them alarmed at the staggering increase in coal pollution that would result from the nuclear shutdown. The union said it would be a lot cheaper and more effective to simply replace nuclear management; the workers were competent, but had been led by managers who weren't. Others used the committee hearings to goad the government into action on deregulation and privatization, saying: "What's the point of talking about this if we don't even know what's going to happen to Ontario's marketplace?" In October, the committee recommended the government take a closer look at the Andognini plan to determine if it really was necessary to shut down all those nuclear units. That never happened.

There were some suggestions at the time that the Americans had recommended the shutdowns in order to make it easier for them to quickly improve performance of the newer reactors and thus ensure that their reputations as turnaround specialists would be burnished. They had strongly rejected the union's position that it was essentially a management problem and that there was no need to shut down 5,000 megawatts of capacity that, despite its problems, was still being operated safely. We will never know who was right, the PWU or the Americans, but the program added about $6 billion to the Hydro long-term debt and greatly increased fossil emissions. It also led to the power shortages in the summer of 2002 which, in the clumsily deregulated marketplace of the Eves Conservative government, inevitably led to huge rate increases.

It had been known since before the time the Energy Competition Act was passed in 1998 that the nuclear recovery program was in trouble. True to form, the Conservatives stubbornly denied any mistake in their approach. In Question Period on October 27, 1998, NDP Energy Critic Wayne Lessard addressed Energy Minister Jim Wilson: "There's one thing we do know, and that is that you are paying top dollar for your nuclear asset optimization plan, but from what we heard yesterday, the only conclusion we can reach is that the plan isn't working. Your own staff tell us that the nuclear assets aren't worth a plug nickel. That means only one thing, that rates are going to go up. If the nuclear power plants fail, rates are going to go up."

Wilson's reply: "The plan that the Ministry of Finance and the group of experts put forward and presented to you yesterday indicates that prices will remain stable and indeed go down over the next few years as competition is introduced."

In 2002, the nuclear power plants failed and rates went up.

Carl Andognini left the country in January 2000, taking with him a huge severance package that included a lifetime pension of US $12,500 per *month*. Not bad for three years' work. Andognini's successor, Gene Preston, had been recruited from another publicly owned utility, the Tennessee Valley Authority, where he had been a nuclear plant manager. He left OPG in September 2002 for reasons that were never made public. As this book goes to print, all reactors that were shut down on the advice of these American experts remain idle. Billions of dollars, in addition to the costs of refurbishing the B reactors, have been spent trying to bring back Pickering A's reactors, so far unsuccessfully. Yet another nuclear fiasco. If these older reactors had to be shut down, as Andognini maintained, we should have written them off at the time and spent the money on energy efficiency and renewable power instead. We would have been much further ahead.

In November 1997, feeling the heat from the private sector and perhaps buoyed by the Americans' assurances that the troubled nukes would be humming along nicely by 2000, the government issued its White Paper—*Direction for Change: Charting a Course for Competitive Electricity and Jobs in Ontario*. It promised an open wholesale and retail market by sometime in 2000 and the separation of Ontario Hydro into a single generation company, (a repudiation of Macdonald's plan for at least four entities) and a transmission company.

The White Paper was well-written and almost persuasive in those days before retail competition had happened anywhere in the U.S. There had been no negative experience yet to dampen the enthusiasm for open markets. The document related the inexorable U.S. march towards deregulation, repeated the glowing predictions of power price reductions south of the border and warned against Ontario "falling behind" other jurisdictions that would be more competitive as their costs came down relative to Ontario's. "By failing to act, Ontario could find itself with rates significantly above those in competing jurisdictions, and lose out in the race for investment and jobs."

Rereading the White Paper five years later, one sentence in particular jumps out: "The government is aware of the need to provide as much rate certainty as possible to electricity consumers." It certainly read well at the time.

In June 1998, the third Conservative Energy Minister in as many years, Jim Wilson, introduced the government's long-awaited *Energy Competition Act*. It followed the general restructuring plan of the White Paper but did not commit the government to opening up the marketplace in 2000, a wise hedge, as it turned out. The bill broke up Ontario Hydro into a generating company (later named Ontario Power

Generation), a wires and retail company (Hydro One), and the Independent Electricity Market Operator (IMO). The Ontario Energy Board was given greatly expanded regulatory powers and the 250-plus municipal electric utilities were told to stop being hydro commissions and convert themselves into Ontario business corporations.

As he defended his bill, Energy Minister Jim Wilson brazenly invoked the spirits of Whitney and Beck while defending his dismantling of the public power system they had founded. After praising Hydro's accomplishments over the last nine decades, Wilson offered up this mixed metaphor: "This has come about because of Sir Adam Beck's and the former Conservative Premier James Whitney's vision for Ontario Hydro. This vision has essentially been completed and it's time to open a new chapter in the history of our electricity industry." Actually, Jim, Whitney's vision was that our province's hydropower would *never* be made "the sport and prey of capitalists."

Surprisingly, the bill also contained a remarkable degree of protection for electricity industry workers. According to the *Act*, all utility workers, provincial or municipal, would not lose their jobs, their seniority, their unions, their collective agreements, their wages, their benefits or their pensions, as a result of legislated changes in the utility sector. The message was that workers should not have to pay the price of restructuring if everyone else was benefiting. It read like NDP labour legislation and was one of the two aspects of Bill 35 our party could support (the other was the enhanced authority of the Ontario Energy Board). Was this the same Conservative government that had stripped Ontario's agricultural workers' right to join unions and retroactively decertified those already in place? Was this the government that had declared war on public sector unions and made it more difficult for non-unionized workers to organize? Was this the same Premier who steadfastly refused to meet with the province's labour leaders? Indeed it was. The anomaly was explained by the turnaround in the attitude of the Power Workers' Union towards deregulation and privatization. To the dismay of the rest of the labour movement, particularly his own parent union, the Canadian Union of Public Employees, John Murphy announced that the PWU supported Bill 35 and pledged cooperation with the government and other stakeholders to make restructuring "a smooth transition." Murphy made speeches predicting that many more jobs for electricity system workers would result from an open, competitive marketplace. Local PWU leaders from around the province met and approved the union's new position. It was clearly a coup for the government that the workers themselves were in favour of its plans.

I have the greatest respect for the PWU leadership and have no reason to believe that their support of competition and privatization is not sincere. They believe that all this confusion will settle down and ultimately be good for their members. That hasn't been the experience in the U.S., however. For the sake of the Power Workers' Union, and not just because I once worked for Ontario Hydro's Construction Division, I hope that they turn out to be right and that what has happened to their colleagues in the U.S. does not happen here.

On October 29, 1998, the *Energy Competition Act* was passed by the Conservative majority, 92 years after a previous Conservative majority had passed legislation creating Ontario Hydro. Every single New Democrat in the Legislature stood up in turn to sorrowfully vote against this death sentence for non-profit public power. The Conservatives jeered, smugly savouring a major victory for the Common Sense Revolution.

16
The Ontario Disaster—Part 2

Despite years of warnings that deregulation would lead to power shortages and higher prices, the Eves government forged ahead with what became the world's shortest experiment in retail electricity competition.

"Who could have known?" Premier Eves pleaded frequently in the Legislature, as he struggled to account for the power shortages and high prices that doomed his deregulation experiment in the summer of 2002. Exactly three and a half years after the passage of the *Energy Competition Act*, the Conservatives "opened" Ontario's power market on May 1, some 18 months behind schedule. Less than six months later, the market was in chaos with skyrocketing prices and threatened power shortages dominating the province's headlines. Every Question Period was a barrage of angry accusations for the Premier as he tried to explain why hydro competition was having exactly the reverse effect he and other deregulation proponents had predicted.

On our side of the Legislature, we were soon able to lip-sync along with the Premier as he reliably chanted the trinity of unexpected events that were supposedly causing the problem: "The Pickering nuclear rehabilitation is way behind schedule. One of the Bruce reactors went down unexpectedly. It was the hottest summer on record since 1955." Our only challenge was in trying to guess, on any given day, in which order he would once again offer up these supposedly exculpating events. We would have enjoyed seeing him squirm more had we not known of the hardship and distress his now-disastrous decision was causing all over Ontario.

Sometimes, the Premier would be combative: "Let's see *you* predict the weather!" he would shout across the floor of the Legislature. "*You* tell us when a nuclear unit is going to go down," he angrily challenged.

The message was always the same: don't blame me. Deregulation was supposed to mean suppliers would be flooding the marketplace with cheap power. Who knew that it wouldn't work out that way? Who knew there would be shortages?

Pat McNeill knew. In September 1997, when the Conservative government was putting the finishing touches on its White Paper, the Vice President of Strategic and Investment Planning at Ontario Hydro was explaining to the Toronto Electric Club the two reasons Hydro was giving for spending billions of dollars on the just-announced nuclear recovery program: "Number one," McNeill said, "virtually all current projections for the North American marketplace agree that the continent's generating surplus will shrink considerably in the early years of the next decade. And number two, those with power to sell at that time will likely make excellent returns on their assets." McNeill was right. Those whose business it was to know were nearly all predicting a decline in generating capacity, relative to demand, over the next few years. The analysts only differed on how steep the decline would be and what its effect would be on prices. Ontario Hydro had done its homework. If it could only get its nukes running more reliably, it could make a killing in the U.S. energy marketplace, just as Bill Farlinger had suggested two years earlier.

The *Hamilton Spectator* knew. A few days before McNeill's speech, the newspaper reported that Ontario Hydro was making contingency plans for what could be a significant power shortfall by 2000. "[Hydro's] problem-plagued nuclear reactors make it difficult to estimate how much generating capacity it will have in future years. Current projections show the utility will suffer a shortfall of 800 megawatts late in 2000 if two of four Pickering A nuclear units are slow in returning to service. That shortfall will grow by 300 to 400 megawatts each year Pickering's return is delayed. The implications are enormous." The *Spectator* asked McNeill about this situation. His reassuring words: "It's a sizeable shortfall, but it's manageable."

The North American Electric Reliability Council knew. In 1998, NERC noted that power shortages were becoming more common, a trend it said should be urgently addressed. That was also the year the cracks started to appear in the young U.S. deregulation program, with unprecedented Midwest price spikes as high as US $7,500 per megawatt hour and the first warning signs of shortages and high prices in California. Coincidentally, the Ontario Legislature was debating Bill 35 in Second Reading just as Midwest wholesale prices began their trip to the moon. The Conservatives didn't notice. Neither did the Liberals, who had already voted in favour of the bill.

People in New York, Massachusetts, New Jersey, Pennsylvania, Delaware, Maryland, Arkansas, Louisiana and Texas knew. They went through blackouts in the summer of 1999 that many experts attributed as much to deregulation as to the weather. A North American Electric Reliability Council official warned the U.S. Congress later that year that competition was endangering grid reliability.

Electricity customers in Montana knew. In 2000, their deregulated prices *quadrupled* even though they had more surplus power than any other U.S. state. Where did it all go? It went to California, where the real money was. But there still wasn't enough power to keep the Golden State lit up. For the first time in its history, California was going through persistent brownouts and rolling blackouts, which continued into 2001. The Western U.S. power shortages (largely fabricated by private power companies, it has now been established) severely affected almost every state in the region. The bills for this misadventure will be coming in for at least ten years. Ontario's Energy Minister at the time, Jim Wilson, however, saw no reason for worry. On December 5, 2001, NDP Deputy Leader and Environment Critic, Marilyn Churley, asked Wilson whether he believed that electricity prices would go up in a free market. His answer: "Yes, if demand does outstrip supply we would find ourselves in a California-type situation. Thank God this government is planning ahead and we are not in California." Naturally, we were all relieved to hear that Ontario's electricity system was in the hands of a government that believes in planning for the future.

I knew. In the October 1, 2001, issue of *Hydro Hotwire*, which tracks privatization and deregulation developments in Ontario and elsewhere, I wrote: "I don't believe there is a surplus of power in Ontario. And if we allow power companies to charge what the market will bear, we are sitting ducks for their greed when the weather or plant outages play havoc with either demand or supply or both." Which is exactly what happened the following year. I didn't need a crystal ball to make this prediction. I simply added up the generating capacity of all of Ontario Power Generation's plants, including the Bruce Nuclear station that was then controlled by British Energy. Unless absolutely every watt of available capacity was running flat out, which never happened, there wasn't enough to meet Ontario demand, which had peaked that summer at well over 25,000 MWh.

Even if I had lost my calculator, however, I would have known that there wasn't enough power in Ontario. All I had to do was read the newspapers that summer, when the Independent Market Operator admitted that it had to import power to keep the lights on. Premier Harris, who was about to announce his retirement, was unconcerned. Howard Hampton was "Chicken Little," spreading needless alarm. Things would be fine. Lots of power was there. Lower prices were just around the corner. To demonstrate his conviction, Premier Harris announced in November 2001 that the market would definitely, positively, this time for sure, open up the following May.

Ralph Klein knew. The pugnacious premier of Alberta has, we can be sure, often secretly wished that he had never heard of deregulation. Before it became the first Canadian province to pass a deregulation law—the *Energy Utilities Act, 1995*—Alberta's wholesale power prices were around $14/MWh, among the lowest on the continent. Among the lowest on the planet. "Not good enough!" said Klein, who promised that deregulation would drive these prices down even lower as generators swarmed into the province, eager to sell power at rock-bottom prices. The swarm must have gotten lost somewhere over the Rockies. It never showed up. As demand in Alberta grew, acute shortages developed and brownouts were a constant threat. By 1998, the supply/demand "cushion" had dropped in half, to less than 500 MW, from what it was prior to deregulation. On many days, only a small transmission connection with BC Hydro kept the province from blackouts. When the western U.S. power "shortages" appeared in 2000, Alberta's market prices soared. The province suddenly found it had to compete with California for every watt of electricity British Columbia could produce. Thanks to Ralph Klein's faith in competition, California prices became Alberta prices. In the second half of 2000, wholesale prices were typically *ten times higher* than the pre-deregulation price, often hitting $500, sometimes more. Factory owners started talking about leaving the province. The media started paying attention. Average Albertans began discussing electricity issues for the first time in their lives. In all of North America, only California had higher prices than Alberta.

Alberta was one of those places that staggered the introduction of competition. Wholesale rates had been deregulated in 1996, with the establishment of the Alberta Power Pool. Retail rates for residential, farm and small business customers were to be regulated by the Alberta Energy Utilities Board until January 1, 2001. Just because they were regulated, however, doesn't mean they escaped rising wholesale prices. Local utilities applied for and were regularly granted the right to pass on the higher costs of wholesale power to their retail customers. Between June and October of 2000, retail rates went from 5¢/kWh to as high as 25¢/kWh. As retail deregulation day approached, no one was looking forward to the privilege of choosing their electricity supplier. The regulated rates were high enough, thank you very much. The Klein government had a very large issue on its hands, one that reached into every household. Both opposition parties were demanding a fall election so that they could victoriously cancel deregulation before the unbelievably high wholesale power rates hit retail customers full force in January. Klein disappointed them. The election would not be held until mid-March. In the intervening months, he scrambled to paper over, with taxpayers' money, the awful truth.

In September, the government announced that it would compensate Albertans for higher power bills and "put an extra $300 in the pocket of every Albertan 16 years of age and over." The money would be paid in two tax-free installments: $150 in November, a Christmas present, and $150 the following April. That second payment would be sent out—coincidentally, I'm sure—about three to four weeks after the election. At the same time, beginning in January 2001, all residential customers were to be given a $20 per month rebate on their power bills for one year. The rebates alone, however, would not be enough to protect Alberta power consumers from rising wholesale prices, so on November 30, the government ordered a small customer price cap of 5 cents per kilowatt-hour. Rebates *plus* price caps. The cap was called the "Regulated Rate Option," to promote the illusion that deregulation was still alive. The power industry howled, of course, saying it would go under in weeks if prices were capped so low, even though the regulated utilities in neighbouring provinces and states were able to sell at that rate, or less. Three weeks later, though, the price cap was obligingly increased 120 percent, to 11 cents. In compensation to confused, concerned consumers, the monthly rebate was doubled to $40. Finally, in mid-January, additional rebates, called "Market Transition Credits," were given to small businesses. Klein was frantically reaching into the provincial treasury and throwing money in all directions to get through the next election.

It all worked. Two months later, Alberta Conservatives won 74 of 83 seats, their largest majority in 20 years. And all for a mere $2.3 billion, which is what the rebates ended up costing. Most of that money went to large industrial and commercial users. The Alberta rebates ended in December, 2001. In 2002, wholesale prices settled down somewhat, but still averaged many times the pre-deregulation price. On October 31, for example, on the 17th anniversary of the Halloween Accord that deregulated Alberta natural gas prices, the wholesale price of electricity in the province averaged nearly $53/MWh, almost four times the previously-regulated price. The day before, the average was $66. Two days before that, prices came in at $87. In fact, during that last week of

October 2002, Alberta's power prices were actually higher than Ontario's, no small feat. The 2002 price cap—excuse me, the regulated rate option—was as high as 6.5¢/kWh, much greater than the pre-deregulation price. Some new generation has come on-line in Alberta in recent months. More is promised, but most observers say it won't be enough to put much of a dent in prices. The voters will have another shot at Klein in 2005. I wonder what he'll come up with then?

How does a province that is floating on a sea of energy come to have the highest electricity prices in Canada? For the same reason that Montana, just south of Alberta, is also paying very high prices despite an abundance of generation capacity. Deregulation is the reason. No other explanation makes sense. Every province and state touching Alberta and Montana is still regulated and every one of them has lower power prices. *Much* lower prices. This is clearly not because Saskatchewan, Idaho and North Dakota have better weather. Mother Nature is not to blame for the sad situation in Montana and Alberta, and now in Ontario. It sickened me every time I heard Premier Eves blame Ontario's high deregulated prices on the weather. Let's be honest, Premier: power prices that reach thousands of dollars per megawatt hour are not an act of God. They're an act of greed, the greed that comes with Hydro privatization and deregulation.

It seems obvious to most of us now that Klein's faith in private sector competition was woefully misplaced. Many free market ideologues still say we shouldn't give up. Stick to your guns. It *will* work. Have more faith. Nancy Southern, for example, runs the power generation division of Alberta's Atco Group, one of the beneficiaries of Klein's policies. She said it's all about "growing pains." Here's her helpful explanation of the problem: "I think whenever you make a major transition from a regulated environment to a deregulated environment... you're going to experience some immature happenings and what has to be a maturing of a transparent system that allows you to gauge supply and demand." Southern, whose company announced in October 2002 that it would build a natural gas plant in Ontario, also claimed that the power situation is actually much better in Alberta these days, even if consumers' bills aren't lower. "I think that's positive," she said, "and I believe that the same will happen in Ontario, but it's a matter of two to three years before the market establishes itself and can identify the supply signals and demand signals appropriately." Someone needs to remind Southern that it's now been seven years since Alberta wholesale and industrial prices were deregulated. Shouldn't this be working by now, even by her standards?

I'll allow that there's a *chance* that Alberta's wholesale power rates will ultimately be lower than what they were under the former regulated system. That chance is zero. Even after adding in the effects of fuel cost increases and other inflation factors since 1996 to the former $14/MWh price, there is no way that power producers, without a stable customer base and guaranteed rates of return, would be interested in such a marketplace. Don't take my word for it. Listen to Stanley Marshall, the head of one of Canada's larger private power producers, Newfoundland-based Fortis Inc. Speaking of Alberta, he said that the signal for private power companies to invest in new electricity generation is high prices. He said this in November 2002, when prices averaged over $60/Mwh. Is this not high enough?

Back in Ontario, with its "surprising" power shortages and high prices in the wake of deregulation, there should have been no surprise. From at least 1997 onwards, there were warnings that Ontario *could* be short of power and there was ample evidence from around the continent that deregulation was no magic wand. Still, the Conservatives stubbornly held to the opposite view. In May 2002, two weeks after market opening, I asked the new energy minister, Chris Stockwell, about a confidential study by the Independent Market Operator that seemed to suggest higher American power prices were coming to Ontario. His cocky, dismissive reply: "The difference between those American markets that were undersupplied and the Ontario market is that we are oversupplied. Because we're oversupplied with power, we will not have the same kind of spikes that will drive up prices in that range." Three months later, the Independent Market Operator was, at times, importing nearly 4,000 MW to prevent blackouts. Average Ontario wholesale prices were *double* what they were when Stockwell had declared himself unconcerned.

Stockwell also joined his predecessor, Jim Wilson, in conveniently forgetting that the Conservatives had, in the past, effusively praised Alberta and California's deregulation. In the Legislature on June 4, 2002, he smirkily told me: "Let me explain California to the people out there and to both friends opposite. California was a situation where they had a Democratic governor. Democrats are a lot like Liberals and the NDP. This Democratic governor—his name was Gray Davis—decided he was going to privatize the energy sector in California."

In fact, it was Republican governor Pete Wilson who signed the 1996 deregulation bill in California. Moreover, privatization was not an issue there, as most of California's generation is investor-owned. Stockwell was perhaps thinking of the Gray Davis who *cancelled* deregulation in 2001. Three weeks later, however, Stockwell reasserted his contention that the problem in California was due to Democrats. I wondered aloud: who was briefing this guy? In August, Stockwell was replaced as Energy Minister. We miss his incisive insights into complex power system issues.

Admittedly, the first two months of deregulation looked good for the Conservatives, with weekly weighted average prices sometimes as low as 2.6¢/kWh. By the end of June, the average price since market opening was 3.5¢/kWh, nearly 20 percent lower than the benchmark 4.3¢/kWh that the Ontario Energy Board had set as the estimated average wholesale rate for the first year. The honeymoon was soon over. In the first week of July, prices rose sharply to 5.4¢/kWh. By mid-July they were 6.8¢/kWh, by mid-August, 8.2¢/kWh, and by mid-September 8.7¢/kWh, at which point they started to moderate somewhat. At the end of October, the average weighted price of power since market opening stood at 5.6¢/kWh, 30 percent higher than the 4.3 benchmark price. That's 5.6 cents for *every* kilowatt hour purchased since May 1. If that rate were sustained, the province's annual hydro bill would go up by about $2 billion in the very first year of deregulation. Even though the hottest summer since 1955 was over, Eves was sweating.

We didn't have to wait until the end of October to learn that this whole thing was turning into a disaster. I had been travelling the province in the NDP's Public Power Bus throughout much of the spring and summer to get firsthand reports from those who paid the hydro bills.

In Wawa, 150 kilometres north of Sault Ste. Marie, residents and businesses alike were stunned when they opened their bills. Roger Guindon, a father on a fixed disability pension, told me his latest hydro bill was double what it had been a year earlier. Resident Helene Scott saw her latest home hydro bill spike from $165 to $347. River Gold Mines and Dubreuil Forest Products, the region's two major employers, told me they were paying 50 percent more for their power. Both companies warned that if hydro prices remained this high, they might be forced to cut staff or even consider shutting down their operations. Several local businesses, especially those that could not cut back on their hydro use and still stay in business—grocery stores and hotels, for example—said the higher hydro costs would be the death of them. Municipal officials worriedly reported they'd have to cut services or hike taxes to cover the 44 percent power price jumps they were being forced to pay. "Hydro deregulation is pushing our businesses and retirees to the edge so they can no longer live in this community,"

The Public Power Bus travelled the province in 2002. In every region, I found that the people of Ontario are as committed to a regulated non-profit hydro system today as they were a century ago. Almost everyone I spoke with, in person or on the phone, was opposed to privatization and deregulation, especially when they saw their hydro bills double in the disastrous summer of deregulation. *Ontario NDP Caucus Services*

Councillor Howard Whent told me.

In the Hamilton-Niagara region, 650 kilometres southeast of Wawa, the stories were the same. Dave Rihbany, a lifelong Conservative supporter, invited our bus to his fish market so that he could show me his "rip-off" hydro bills, which had gone up 70 percent in only one month, from July to August. "I challenge Ernie Eves to get out there and talk to real businessmen," Rihbany said. "Things aren't wonderful. The Conservatives have lost my vote." Down the road in Welland, Maria and Pasquale Ramundo have run a small deli for 17 years. After receiving a hydro bill in September that had nearly doubled from the previous year, they wondered if their hard work building the family business had been in vain. "Something's got to be done," said Pasquale, "otherwise, we can't survive anymore. I wanted to pass the business on to one of my sons. But if it keeps going like this, I'm just going to turn the key and lock the door."

Everywhere the Public Power Bus went—a fish and chips restaurant in Orillia, an Asian food market in Rexdale, the dairy farm in Woodbridge, and all the Tim Horton's stops along the way—people were distressed and outraged over skyrocketing deregulated hydro prices. I brought many of these hydro horror stories into the Legislature. At first, the Premier waved them away. "Prices go up, prices go down, all that counts is the average," sums up his standard response. After all, hadn't prices been lower in May and June? They'll go down again, you'll see. But they didn't, not by much, anyway. News of growing consumer distress began to push its way into the Premier's Office via his own caucus. In the Legislature, Premier Eves shifted ground from minute to minute. He blamed the weather. He whined about the nuclear units still in the shop for repair. He expressed sympathy for the poor consumers—feeling their pain. He attacked the opposition parties, particularly the NDP, as a worse alternative to his government, frequently jabbing his finger in my direction as he vowed that he wouldn't "pull out the Hydro credit card like you guys [the NDP] did, and rack up another $3 billion in debt." He was, of course, referring to the additional long-term Ontario Hydro debt that showed up during our term because that was when Darlington began operating and therefore the station's full costs had to be shown on Ontario Hydro's books.

There was only one thing that Premier Eves and I agreed on when it came to deregulation and privatization: we both were disgusted by the hypocrisy of the Liberals, who had been consistently in favour of selling off our public power system and deregulating prices, right alongside the Conservatives, every step of the way. Liberal leader

Dalton McGuinty had, in fact, recently been out fundraising on Bay Street, giving written promises that the Liberals believed in both deregulation and privatization. Now, they were furiously backpedalling, trying to distance themselves from this policy debacle. Liberals have always been good at reading the polls. Some people say that's *all* they're good at, but I personally would not be so unkind.

When Premier Eves wasn't in the Legislature to take the questions, his latest Energy Minister, the soldierly John Baird, kept up a solid front

---

**ENERGY SECTOR RECEPTION
FOR
DALTON McGUINTY**

RECEPTION COMMITTEE

Richard King
Power Budd LLP
Co-Chair

Sean Conway, M.P.P.
Renfrew-Nippising-Pembrooke
Co-Chair

Frank Carnevale
City Hall Group

Jan Carr
Barker, Dunn & Rossi

Dennis Fotinos
Enwave

Charles Keizer
Power Budd LLP

Danielle Kotras
Hill & Knowlton

October 31, 2001

0 8 NOV 2001

Dear

    We are writing to invite you to a reception to meet Dalton McGuinty, Leader of the Official Opposition. Dalton was elected as M.P.P. for Ottawa South in 1990 and Leader of the Ontario Liberal Party in 1996. Prior to becoming Leader, Dalton was the Opposition Energy Critic from September 1990 to July 1993. Throughout Ontario's electricity restructuring process, Dalton and the Ontario Liberals have been consistent supporters of the move to an open electricity market in Ontario.

    We hope that you will join us and other friends at a reception on Wednesday, November 28, 2001 from 6:00 p.m. to 8:00 p.m. in the Great Hall at the Hockey Hall of Fame, BCE Place, 30 Yonge Street, Toronto. Tickets are $350 per person, for which a tax credit receipt will be provided. To reserve your ticket, simply mail or fax back the attached reply sheet to (416)260-1606, or contact Jennifer Simpson at (416)260-1608 ext. 224 to confirm your attendance.

    We look forward to seeing you at the reception.

Yours truly,

Richard King
Power Budd LLP

Sean Conway
M.P.P., Renfrew-Nipissing-Pembroke

Ontario Liberal Fund
243 College Street Suite 401 Toronto, Ontario M5T 1R5
Telephone : (416)260-1608 Fax: (416) 260-1606

Liberal Leader Dalton McGuinty sold $350 tickets to Bay Street lawyers and financiers in late 2001 with the assurance that "Throughout Ontario's electricity restructuring process, Dalton and the Ontario Liberals have been consistent supporters of the move to an open electricity market in Ontario."

of rapid-fire optimism. By this time, the spring and summer of 2002, the Conservatives were naming a new Energy Minister every six months because the previous one no longer had any credibility. "There's plenty of power, Mr. Speaker. Prices will go down, Mr. Speaker. On this side of the House, we care about consumers, Mr. Speaker." One of Baird's favourite lines: "The Leader of the Opposition is a 'Howard-come lately,' Mr. Speaker" was clever but not true. McGuinty still thought deregulation could be fixed. In fact he implied frequently that the Liberals would "do a better job of deregulation." Then, around Halloween, Baird began talking about a plan that he was "working on night and day," at the behest of the Premier, to address the problems that the government had been so fervently denying for months. For two weeks, we heard about this impending plan every day in the Legislature. Both opposition parties wanted to know what was in the plan. We'd just have to wait and see, Baird said. But make no mistake about it, he promised, he wouldn't be pulling out that tattered old Hydro credit card again, like some people he could name.

On November 11, 2002, Ernie Eves pulled out the Hydro credit card. At a clumsily staged late afternoon press conference in a middle-class kitchen in Mississauga, a shirt-sleeved Ernie Eves announced a four-year rebate/price cap scheme that must have surprised even the generous-to-a-fault-with-your-money Ralph Klein. Eves's rate cap was 4.3¢/kWh and the rebate was for 100 percent of all hydro energy charges above that mark, retroactive to the May 1 market opening, and continuing *until 2006* for those who had signed contracts above that rate, which was 25 percent of the province's residential customers. No one would pay more than 4.3¢ until "at least" 2006. Those who had already been paying more would get a $75 rebate cheque, before Christmas, naturally. The rest of the rebate would be a credit on future power bills.

In my response to the media I was blunt: "This is a desperate government trying to cover up the high cost of deregulated, privatized hydroelectricity until after the next election. It's a government that is trying to bribe people with their own money." I predicted the rate cap would last until two or three months after the election. If either the Conservatives or Liberals were elected, hydro privatization and deregulation would continue and consumers' hydro bills would go through the roof again.

And where would all this rebate money come from? Since market opening, it seems, Ontario Power Generation had been putting aside some rebate money it would have had to pay out to customers anyway under the terms of a 1998 market power mitigation agreement with the government. There was plenty of cash in *that* pot to cover the rebates, Eves claimed. Problem solved. The Premier then hopped on a plane to Arizona for a week, leaving behind a startled province.

More than six years in the making, the Ontario experiment with retail electricity market deregulation was now the shortest on record. Indeed, with the retroactive price cap, it hadn't even lasted a day.

Reaction to the Premier's retreat was swift and negative. Consumers instantly knew they were being bribed with their own money, and said so on every television news show that asked people in the street what they thought of Eves's plan. The real anger, however, came from the private power industry. "Et tu, Ernie?", they cried disbelievingly. Peter Budd, the province's best-known energy lawyer, also chairman of the Ontario Energy Association, said publicly that imposing a price cap would make investing in Ontario unappealing. Privately, he fumed that the government had consulted no one in the industry before embarking on this reckless path. (Budd's public reaction to Eves's cancellation of the Hydro One sale two months later was more pragmatic, not to mention revealing. He viewed it as "more of a delay rather than a permanent cancellation… Who needs controversy with an election coming?")

Big industrial power users were also perplexed. "Let's face it, in the short term, [the price cap] might look okay," allowed Arthur Dickinson, long-time head of the Association of Major Power Consumers in Ontario. "But how are we going to get new supply on stream? It's beyond me." Smaller businesses were equally critical. Judith Andrew of the Canadian Federation of Independent Business asked: "How are we going to get more generation on stream, and eventually make the long-term better in terms of having lots of competitors out there?" The plan that Energy Minister Baird had been working on all these weeks was declared an instant flop. If it had been a Broadway play, it would have closed right after the curtain rose on opening night.

We learned the next day, however, that Baird had not been burning all that midnight oil just to come up with rebates and price caps. There was even more money involved. "With Niagara Falls thundering in the background," read the florid government press release, Baird declared that "immediate and decisive action" would be taken to increase the province's electricity supply. For the most part, the decisive action meant pulling out the Province's credit card and exempting from taxes any corporation that would henceforth build alternative (natural gas) and renewable energy source generation. Nor would any taxes be levied against revenues derived from the sale of that power. Even the materials that went into the building of new supply would not be taxed.

The "renewable energy" was in Baird's announcement for cosmetic political reasons. The tax breaks were aimed at the more marginally economic natural gas cogeneration sector. With 100 percent investment write-offs and no taxes to pay, natural gas developers *might* be attracted by this deal, especially if they knew that the 4.3 cents was only in place until 2006. (They won't have to wait that long if the Conservatives or Liberals win the next election. With deregulated wholesale electricity still sky high, subsidizing retail rates at this level will quickly become an unsustainable drain on provincial revenues and will be cancelled or phased out quickly following an election).

The Conservatives' cunning plan, in a nutshell, was to have Ontario taxpayers substantially finance the construction of private power stations, but not own them, or even hold shares in them. After all, having the public involved would just undermine the "market discipline" the Eves government had brought to Ontario's electricity industry.

Even as he flourished unprecedented, publicly financed bribes to private companies, Energy Minister Baird refused to let go of the cherished fiction that Ontario had plenty of power. "The province has a sufficient supply of electricity to meet current demand," he shouted to reporters over the thundering Falls. Not far away, however, at the Ontario-New York transmission link, power was flowing in from the U.S. That week, according to the Independent Market Operator, Ontario's average daily net import schedule was just under 1,000 megawatts. This is more than the amount of power produced by the Sir Adam Beck 1, the great Queenston-Chippawa project that will, if the Conservatives cling to power, be ultimately sold to private interests. Because as we all know, the private sector does these sorts of things so much better.

17
The Inevitable Failure of Deregulation

The unique nature of electricity, combined with basic principles of prudent investment, make it impossible for deregulated markets to sustain lower and more stable average power prices than regulated markets.

What went wrong with deregulation? Why has it not worked as advertised, here in Ontario or anywhere else for that matter? Why, after six years of planning involving some of the best minds in the power industry, are we not seeing the abundance of cheap power that Conservative Premiers Harris and Eves assured us would be the natural consequence of an unfettered marketplace? Why have prices gone up, not down?

In Ontario, the problems are being hidden for the time being by the Conservative government in exactly the same way they have been covered up in other places where deregulation has failed: by temporarily returning to regulation and other forms of government intervention while insisting all the while that the idea of deregulation is still sound. Here's what Premier Eves wants to hear from Ontario voters: "Good idea, Ernie, that deregulation stuff. Darn shame about the weather, though. I guess that ruined it, eh? Oh well, not your fault. Glad you froze the rates. Carry on."

Power industry experts privately guffaw at the weather excuse because electricity systems and markets have to assume there will occasionally be weird weather, and must be designed to accommodate it. If you haven't done this, you simply haven't done your job and the problem is entirely your fault. If, on the other hand, you are shopping for more sophisticated explanations of deregulation's failures, there are plenty available from the true believers, those who continue to support deregulation even as it collapses around them. There are basically three schools of thought.

The first is the "It *is* working, just give it time" school. The school motto: "Send the right price signals! Let the market sort it out!" The theory holds that high prices and shortages are exactly what we need to attract new power producers into the marketplace. We wave our money at them and the invisible hand of lust for wealth will guide capital unerringly—in time, of course—to a sustainable balance of private profit and public good. If the demand for power is higher than supply and profits go up accordingly, we can be quite sure investors will behave accordingly and come running to cash in on the situation. Then, at some point, supply will exceed demand and prices will naturally come down. As that happens, producers will be forced to drive their own costs down in order to maintain their competitiveness, their market share and, most important of all, their profitability. Those who do this well will survive and flourish, those who do it poorly will lose money and leave. Consumers, of course, can never lose because they will always have a choice and will reliably choose the lower cost, better service, power providers.

A corollary to this invisible hand theory is that consumers also respond to price signals by changing *their* behaviour, which also affects the supply/demand balance. The theory says that high power prices encourage conservation and energy efficiency, thus lowering demand, which in turn puts downward pressure on prices. High prices also encourage self-generation, which increases and diversifies supply. In either case, commercial power producers are further forced to be more efficient and innovative in order to maintain market share and profitability. Or so goes the theory.

The theory of the second school is: "It *will* work, but only if we get the rules right." School motto: "A level playing field in a transparent market." This school also believes in the invisible hand, but says the complexity of electricity systems is such that a visible hand is needed as well: a set of rules that applies equitably to all players and, equally important, ensures the efficient transfer of price signals. The rules can cover a very broad range of market activity: how generators bid into the spot market, access to information on the state of the electricity system, metering and billing protocols and literally hundreds of other aspects of the marketplace. Getting the rules right is critical. A poorly designed market with inadequate rules will restrain, even paralyze, the invisible hand and prevent the development of a sustainable balance between public and private interests.

The third school of thought accepts the theories of the other two as quite true, but adds a physical dimension: the "Inadequate

Infrastructure" theory. School cheer: "More Transmission Towers!" According to this theory, we have an antiquated, patchwork transmission network in North America that has been cobbled together over the last hundred years by scores of now-defunct regional monopolies. The various "patches" were originally designed to move power from domestic generators to users within state or provincial boundaries. They were built to carry only so much power, like a two-lane highway that can carry only so much traffic. Over time, some interregional ties were built, such as the ones that connect Ontario's power system with those of neighbouring states and provinces. But most of these "interties" were built mainly for system reliability purposes, so that neighbouring regions could help out one another with unexpected or seasonal power shortages. They were not designed as 16-lane highways to carry a steady, heavy stream of transcontinental traffic 24 hours a day. The inadequacy of this physical network greatly inhibits power trading and discourages investment in new generation. "If I'm going to put $500 million into a new power plant," the tower mavens say, "I want to have the widest market reach possible. I'm not going to locate my plant where the network severely limits my access to customers. If you want me to invest, build an adequate infrastructure, build more transmission lines, many more."

All three theories make perfect sense as thought experiments. They appeal—dare I say it?—to common sense. Of *course* competition encourages efficiency and innovation. Of *course* you need equitable and transparent rules to attract new market entrants so that competition will be vigorous. And of *course* you need to know you can get your goods to market before you invest in a production facility. The apparent truth of these theories explains why the idea of open electricity markets has been so seductive, and continues to be believed in the face of the obvious reality that deregulation is not working.

The school of hard knocks, however, suggests that common sense is wrong in this case. Our experience to date is that either the idea of deregulation *is* wrong or we're simply not capable of executing it properly. Either way, it's been a huge mistake, one of the biggest economic blunders of our time. How did it happen? I think I know, because I have seen the same thing happen all the time in the Legislature over the last eight years of Conservative rule. It is the triumph of ideology over experience and, in the case of electricity, over physical reality itself. So unshakable is the Conservatives' faith that privatization and deregulation are *always* good for us that they simply avoided coming to terms with two facts about electricity that textbook supply and demand theories can't accommodate. First fact: electricity cannot be stored. It must be produced at the very instant it is consumed. When you switch on a light, you are using power that is being generated right at that moment. This physical reality of the nature of electricity—its un-storability—has profound implications for electricity systems. Second fact: electricity demand is highly variable, on a daily and seasonal basis. This is especially true in a place like Ontario, with our climatic extremes.

To illustrate the challenge that power systems face from the combination of these two facts, let's use a real-life example of variable Ontario demand that happened somewhere near the mid-point of our brief deregulation era. At about 3 a.m. on July 29, 2002, total demand in the province was at its lowest for that day, about 16,000 megawatts. The spot market price of power was just over $40/MWh. This minimum demand is the "base load" of Ontario's electricity system, the amount of power that will be needed all the time. Nuclear stations are typically used to meet Ontario base load demand because they are always on (in theory). It takes about a week to safely start or shut down a nuclear reactor. Not much flexibility there. Hydroelectric stations, especially our large ones, are also used for base load power. In recent years, due to the nuclear shutdowns, coal stations have been used for base load as well, chugging away 24 hours a day.

As dawn broke on that hot July day, demand began to rise steadily, reaching a "peak" at around 3 p.m. of nearly 25,000 MW. To meet that increasing demand, more generating units—peaking plants—were progressively called on throughout the day. Some peak demand can be met by hydroelectric stations, but not much. All our Hydro plants together can only supply about 25 percent of Ontario's average daily demand. Most peak power is supplied by our fossil plants because they can be turned on and off on short notice, within minutes in fact. At 3 p.m., every available generator in the province was running flat out, but it still wasn't enough. Nearly 3,500 megawatts, a Darlington equivalent, had to be imported at one point. Ontario spot market prices hit $200/MWh. The Independent Market Operator issued a "power advisory," asking consumers to curb electricity consumption.

Demand then dropped off steadily for the next 12 hours, in a pattern that is as predictable as night following day. By 3 a.m. on July 30, demand was again around 16,000 MW. In the course of 12 hours on either side of the peak, the spread in total provincial demand was 9,000 MW. That is the generating capacity of nearly *three* Darlington-sized nuclear stations (or three coal-burning plants the size of Nanticoke, on the north shore of Lake Erie). The difference is even more dramatic if we look at the difference between the lowest and highest demand over the course of an entire year, reaching as high as 13,000 MW, or four

Darlingtons.

Here's the challenge: Ontario's average daily demand over the course of a year may be somewhere around 17,000 MW. But if that's all the generating capacity we had access to, we'd have brownouts or blackouts whenever demand exceeds this average, which is most weekday afternoons and early evenings. To ensure reliability, therefore, we need a fair bit of generating capacity that is idle much of the time. In fact, we need generation that we might use only a few days a year, like on that hot July day or on a deep-freeze day in February. A related issue is the fact that generating units must occasionally be shut down for maintenance. They are also subject to unplanned outages; nuclear units are particularly notorious in this regard. To cover these outages, we need access to even *more* capacity than is required to meet actual peak demand, so-called "surplus" capacity. Most systems plan for a surplus of around 15 percent over peak demand. This is our reliability insurance.

Why is all this a challenge? Because idle capacity has to be paid for, even when it's not being used. Insurance is not free. Because of the capital-intensive nature of electricity generation, the cost of a kilowatt of power from a station that is used intermittently is more expensive than a kilowatt from a station that is running all the time. Peak power costs more. If we could store power that we generated in the middle of the night from base load stations for use later in the day, we might only need the equivalent of one Darlington nuclear station, instead of three or four, to bridge our peaks and valleys. But we can't do this, not in any meaningful way, which means that a lot of generating capacity has the job of just sitting there, waiting to be called on. Like firefighters waiting in their station for the alarm bell.

Let's now look at the big difference—actually the *huge* difference—between how regulated and deregulated systems handle this challenge of having to pay for idle generating capacity. In Ontario's formerly regulated system, we *averaged out* the cost of power from base load and peaking stations. We spread out the cost of our reliability insurance over the entire day. At 3 a.m. we pay more than the actual cost of power we're using at that time, but at 3 p.m., we pay less. The total costs of the whole system are shared equitably by all customers through a rate structure that is typically set for a year at a time. The price of power in April is the same as in August.

In a deregulated system, by contrast, every generating station is a separate business entity that must stand on its own two financial feet. A peaking station that is needed only a few days a year must charge an enormous amount of money, per kilowatt-hour, in order to cover its construction and standby costs. When power from a peaking station is needed in a deregulated system, it is naturally going to charge as much as the market will bear. Those are the rules. If that means charging $2,000/MWh, as sometimes happened in the summer of 2002 in Ontario, then so be it. A station's got to make a living, you know.

Imagine if we approached firefighting this way. We decide we're paying firefighters too much to just sit around and wait for something bad to happen. We send them home with no pay when we don't need them and let them charge us what the market will bear when we do need them. A fire breaks out, your factory is burning. You call a firefighter for help, who says: "How much is it worth to you to have me come and put out your fire? I'm worth $2,000 an hour. What do you say?" How much would you be prepared to dicker?

Price Signal theorists have no problem with $2,000/MWh power prices. "That's *exactly* why deregulation makes sense," they say. With averaged rates, consumers are not made aware of how expensive air conditioning is during a heat wave, especially if everybody else has the same selfish desire to keep cool. Not sending consumers clear price signals discourages conservation and energy efficiency, which then requires the building of even more capacity. Remember the consequences of the "Live Better Electrically" campaigns? Don't we *want* customers to be motivated to reduce consumption, especially at peak times?

Of course we do, but volatile deregulated prices are a terribly blunt and unsustainable way to achieve this. Here's why: In an hourly spot market, you don't know exactly how much your power is going to cost until *after* you use it. You certainly don't know how much it will cost tomorrow or the next day. As Premier Eves would say: "Let's see *you* predict the weather! *You* tell us when a nuclear unit is going to go down unexpectedly!" How can your behaviour be affected by something that has not yet happened? If, however, you were aware in advance that prices were going to be exceptionally high on a particular day, or at a particular time of day, you would be more likely to plan to reduce your consumption. If, as a matter of public policy, we want to vary prices throughout the day to encourage conservation, we could do that better through a *regulated* system in which actual power costs were publicly posted far enough in advance to have an impact on behaviour. Perhaps it could be on the radio and TV news: "Tomorrow, your hydro is going to cost 25¢/kWh from noon to 6 p.m. Save money, don't overdo the air conditioning." That would be a better way to send price signals to people instead of waiting to hit them in the monthly bill, long after the fact.

I would not, however, favour such a "fine-tuning" approach. I do not want to send hourly price signals to schools and hospitals and nursing homes, the unemployed, women's shelters and your home. Nor do

I want Ontario industries to be constantly calculating whether it is cheaper to send workers home than it is to keep the factory or mill operating. I do not want a society in which the good people of Rosedale, one of Toronto's very high-income neighbourhoods, can keep cool on sweltering summer days in their centrally air-conditioned homes while the good people of nearby Regent Park, with their much lower incomes, cannot afford to turn on the single window air conditioner in their one-bedroom apartments. And I especially do not *ever* want to see prepaid metering in Ontario, as some half-crazed Price Signallers are proposing. Prepaid electricity metering is the latest rage in the U.S. (and in many developing countries) because it eliminates all business risk for the utility. When the old and the poor run out of money, they run out of electricity, no matter what time of the year it may be. It also eliminates the utilities' public relations problems. They no longer have to actively turn off the electricity. It's not because of their greedy cold hearts that people are freezing in the dark. It's because the meter ran out of money.

I am in no way opposed to people understanding the true cost of the power they use, including, perhaps especially, its true environmental cost. It is important that people conserve energy. But we know that conservation has its limits. Energy efficiency is clearly the way to go if we want to reduce demand. And you cannot as easily promote energy efficiency in a marketplace with prices that vary from day to day and from hour to hour. In fact, volatile prices make it *impossible* to calculate the energy cost savings of energy efficiency measures. How much money will you save if you buy that energy efficient refrigerator? In a deregulated marketplace, you only find that out after the bill comes in because you can't know in advance how much you're actually paying for power from hour to hour. In a regulated marketplace, by contrast, this is a fairly simple calculation because the cost of power is stable. So despite what the Price Signal theorists say, volatile prices can actually discourage small-scale investments in energy efficiency. If you don't know whether your savings will be big or small, why bother?

In my view, we should not be relying on high prices to drive our efforts to reduce consumption. Energy efficiency makes long-term economic sense whether the price of power today is 4.3¢ or 8.6¢/kWh because in either case it reduces the need for more generating capacity. Moreover, reducing energy consumption *always* makes environmental sense, whatever the current cost of power. Permanent reductions in energy use can be better achieved through targeted, community based energy efficiency programs for homes and small businesses, and provincially run programs for larger industries, than through the unpredictable workings of an unfathomable (for most people) marketplace. This is exactly what the NDP government was doing years ago, with great success.

The Eves Conservative government has belatedly discovered the virtues of residential energy efficiency. In November 2002, they declared a one-year provincial sales tax holiday on the purchase of new energy efficient home appliances. I know that every little bit helps, but this feeble, scattergun approach will not have much effect because it requires the initial outlay of hundreds of dollars. It would have far more impact if the appliances could be financed at low rates on the hydro bill. But at least it shows that the Conservatives, in time, may actually start thinking about real energy efficiency. In announcing the program, Conservative Energy Minister John Baird told the press: "If we can encourage people to get rid of that 30-year-old fridge in the basement where they put leftovers, that would certainly have a huge impact." What a great idea, John, but why is your incentive so feeble?

If price signals are crude tools for encouraging energy efficiency, they are even less useful for encouraging the construction of new generating capacity in tight supply markets. The Price Signal theorists may protest, but they will have to explain why there is so little new capacity being built these days in deregulated markets. Alberta has been deregulated now for seven years and it was known back in the mid-1990s that the province was heading for shortages. U.S. deregulated jurisdictions have had the same trouble with reluctant investors, who have become even more reluctant in the last couple of years.

So why did the Eves Conservative government, in November 2002, have to offer tens if not hundreds of millions of taxpayers' dollars to the private power industry to build new plants here after claiming for years that the private sector just couldn't wait to invest in more capacity? The main reason, according to Ontario's independent power producers, is the uncertainty surrounding Ontario's nuclear sector. "Sure, there's a shortage now and you're begging us to build," they say, "but will you need us when Pickering A returns to service?"

They're right. In a deregulated, privatized power system why should investors put a lot of money into an uncertain marketplace with no guarantee that they will be able to recover their money? Power generating stations are not travelling carnivals that can be put up overnight on mall parking lots and taken down just as fast. They are long-term commitments that require long-term debt financing. No corporation, no matter how wealthy, would ever write a $500 million cheque to build a generating station. The money is mostly borrowed; banks and other lenders are always part of the deal. The amortization (loan repayment) period for a medium-sized natural gas plant is typically 20 years. Lenders need assurances that they're going to get their money back.

Let's imagine you're a private power developer who wants to put up a natural gas-fired plant in deregulated Ontario. You approach lenders with the proposition. At current lending rates, you know you can produce power for, say, 6¢/kWh. That will be enough to cover your financing and operating costs for the next 20 years, plus the necessary profits, plus the taxes on those profits. The average price of deregulated power in Ontario is now slightly less than 6 cents, but you have a peaking plant, so you can often charge more than that, sometimes much more. Repaying the lenders should be no problem.

Your potential lenders, however, aren't as confident. They have some questions. What impact on prices will this new supply have? After all, as supply increases relative to demand, prices go down, right? Is anyone else planning to build as well? What could your nuclear, coal and hydroelectric competitors sell their power for if they were forced to lower their prices to remain competitive? What will the price of your fuel, natural gas, be in five, ten and fifteen years? Didn't it triple in price only a couple of years ago? Will the transmission links to Manitoba and Quebec be upgraded at anytime during the life of your plant, giving Ontario access to low cost hydroelectric power that you couldn't possibly compete with? Are there any new technologies out there that would make small-scale distributed generation more economically attractive? Will some government, desperately clinging to power, suddenly change the rules and impose a power price cap of, oh, let's say 4.3 cents, which is lower than your cost of production? What will you do then?

I could go on and on with questions like this that careful investors are asking these days. One big reason there is currently so little investment in new generating capacity is that these questions are impossible to answer with any degree of certainty acceptable to investors who are being asked to put up money for long periods of time. They've already been burned by market volatility. Tens of billions of dollars have been lost in the North American private power industry since 2001, when Pacific Gas and Electric went bankrupt, followed by the debacle at Enron and the crashing of several other, once high-flying industry giants. Investors are not lining up to put their money into new supply for a deregulated power industry. It's too risky.

Let's say instead that your proposed generating facility has a 20-year power purchase agreement with a government-owned utility. The agreement calls for 6¢/kWh, with adjustments should the price of gas go significantly up or down. Now your risk is dramatically reduced. The only major risk is that the plant will break down, but you can cover that with insurance and warranties. *Now* your lenders are interested. Guaranteed profits sure sound better than 20 years of uncertainty.

Investors hate uncertainty, everyone knows that. Ernie Eves was a banker before he became Premier, so he should have known that as well. Even if you're the speculative type and you can deal with a high degree of uncertainty, you would expect a *big* payoff in return for investing in such an expensive long-term project as a power plant, the economic viability of which is subject to so many unpredictable variables. This is why reasonably priced money for new power projects is so hard to come by these days. "High volatility? No guarantees? No thanks. Come back with an iron-clad contract from a government-backed utility. Then we'll talk."

You can now appreciate why the Price Signal theorists are wrong when it comes to the power industry. Price isn't the real problem, uncertainty is. Deregulation, by definition, means uncertainty. Therefore, deregulation is the problem. Investors will only plunk down their money if they are reasonably certain that average prices will stay above the costs of production. The panoply of tax breaks now being offered by the Eves government will certainly have the effect of reducing the costs of production, which is meant to make such projects more attractive to lenders. It may work, but please don't call it deregulation. It is massive government intervention, using taxpayers' money to bribe private producers to build a facility that they can then turn around and price gouge us with at the first (and every) opportunity.

If the Price Signal theory is wrong, so is the "Right Rules" theory. Perfect rules may ensure a level playing field, whatever that is, but if the game itself has too many unpredictable variables, it will be hard to find backers for your squad. In a deregulated power marketplace, uncertainty rules. I'm not sure it is humanly possible to come up with the "right" rules. And even if someone did, it would be a daunting task just learning them. The Ontario Market Rules Manual, for example, is 928 pages long. Good luck explaining all these rules to your prospective lenders. And don't forget the Interpretation Bulletins that go along with the rules. Those are extra. Right Rulers cannot explain why their analysis of what went wrong in some jurisdictions is contradicted by what went wrong in others. Here's an example: "California's problem is that its market rules forced all wholesale power sales through the spot market and didn't allow for direct supply contracts between generators and customers." But Ontario rules do allow such contracts, so why did prices go up here as well? Here's another example: "Alberta deregulated the wholesale market first but continued to regulate residential and small business rates. That was their big mistake." Once again, Ontario took a different path and simultaneously opened up the wholesale and retail markets. So what happened here?

Get a bunch of Right Rulers together and the meeting quickly starts to sound like a medieval convention of Ptolemaic scholars going through mental contortions to accommodate astronomical observations that appear to contradict their theory that the Earth is the centre of the universe. The models get progressively more intricate and complex, but they still don't work because the central premise is wrong, as Copernicus eventually explained. So it is with our attempts to find the perfect market rules. The central premise is wrong: competition will not produce a more efficient marketplace that will benefit consumers because of the inherent, unavoidable uncertainty of power markets. No rules we can devise (even if we could enforce them all, which is another problem altogether) will get around the two big facts: we cannot store electricity and demand is variable, so *someone* has to pay for a large amount of idle generating capacity to ensure reliability. Regulation is a way of spreading that cost rationally and equitably. Deregulation is a chaotic daily battle over how much that reliability insurance will cost *today* and who will get stuck paying more than their fair share.

Paradoxically, the less reliability insurance you have in a deregulated system, the more the insurance costs. Another way of saying this is that the tighter the supply of peak power, the more the suppliers can get for it. Deregulation carries with it a built-in incentive for generators to withhold supply from the marketplace so that prices rise, otherwise known as market manipulation and price fixing. This is what happened in California and elsewhere, in Ontario. Deregulation makes lower reliability more profitable for suppliers and more costly for consumers. To put it bluntly, a deregulated market creates incentives for private power generators to manipulate the market, create artificial electricity shortages and then gouge consumers when the price goes up. You can make a rule that forces suppliers to offer all of their available capacity to the marketplace when told to do so by the system operator, but then you're using regulation to make deregulation "work."

That leaves the Inadequate Infrastructure school of thought. Here we have the strongest team in the deregulation league because it's true that we have a patchwork of interconnected transmission systems in North America and it's true that this inhibits power trading. I could cite lots of numbers to demonstrate this but your eyes would quickly glaze over. It's also true that upgrading the transmission system would make more generating facilities available to a wider marketplace, enhancing competition. Give lower-cost generation better access to a wider market and it will quickly drive out higher cost generation. Consumers will win. Not only that, our continent's reserve margins, our reliability insurance, won't have to be as large. If we can ship more power longer distances, we won't need as many coal-burning peaking plants and the ones we do have will be more productive, therefore cheaper. Consumers win again; the environment wins as well. This is starting to sound good. It also seems to make sense when we look at a couple of real-world examples.

Remember Alberta's problem? Alberta is almost an island, electrically speaking, with only one small inter-tie with British Columbia and an even smaller one with Saskatchewan. That relative isolation fully exposed power-scarce Alberta to the extreme volatility in Western U.S. power prices in 2000 and 2001, as well as to the impact of power plant outages within the province. Even one small plant going down could send prices through the roof. If Alberta had been more interconnected, it might have been able to get power from Manitoba, two provinces over, which has lots of water-cheap hydropower to sell.

Then there's Quebec, which makes hundreds of millions of dollars a year selling an enormous amount of power into the northeastern U.S. The wide transmission highways that slash through Quebec for nearly a thousand kilometres, from James Bay to the U.S. border, keep a lot of lights on in New York and New England, for less than the cost of power from natural gas or even coal stations. Is this not a good thing? Why wouldn't we want to do more of this? In fact, if Ontario had more inter-tie capacity, we probably wouldn't have suffered from shortages in the summer of 2002. Deregulation wouldn't have been as discredited. Maybe more towers *is* the answer. This is certainly the opinion of the U.S. Federal Energy Regulatory Commission, which has been trying very hard to restructure that country's transmission sector in ways that will supposedly encourage more investment in expansion and upgrades. The Bush-appointed FERC commissioners are adamant that erecting more transmission towers is the key to making deregulation work.

This way lies madness! We must resist these fantasies because they will, if unchecked, take us deeper into the pit of expensive non-sustainability. I'm not talking about the necessary upgrades to our existing transmission system. We have yet to recover fully from the 1998 ice storm, for example, and sections of Ontario's grid are very old, with towers dating back to Beck's time. As a former tower builder myself, I can't help noticing these still-functioning old-timers as I travel through rural Ontario. They've been wonderfully maintained over the years, as have the transformer stations on the ground that form an equally important part of the network. But they will not last forever and, in any

Left: A double circuit transmission tower spans Rainy Lake Causeway in my home town, Fort Frances. Having helped build this and other towers in the 1970s, I can well imagine what it must have been like, seven decades earlier, to do the job with actual horsepower, right. While the Ontario Grid must be well-maintained and upgraded to ensure sterling reliability, building massive transmission superhighways to trade power with the United States would be a huge economic and environmental mistake. *Hydro One Archives*

event, new technologies are available to reduce line losses (which cost us all money) and make the system more resistant to the forces of nature. There is no question that we must continue to invest in our grid to ensure its continuing reliability and improve its efficiency. We still need the towers and, in some places, we do need more of them. But we do *not* need transmission superhighways that scar our landscape and embroil whole communities in prolonged, expensive battles to prevent them from being built. We can greatly reduce the need for more transmission infrastructure by building generation closer to home, small-scale facilities that can bypass the main grid because they are built to serve a relatively small area. Off-grid power production is commonly called "distributed generation," or DG. A DG facility might be a small natural gas cogeneration plant in the basement of an office tower or hospital complex. It could be a wind turbine such as the one built by the Toronto Renewable Energy Cooperative on the Canadian National Exhibition grounds. It could be a solar panel array on the roof of a suburban school or a local arena. And someday soon, it could be a hydrogen fuel cell in your basement or at the end of your block.

I am not claiming that all of Ontario's future power needs can be met with such small-scale power projects. I believe very strongly that an aggressive and continuous program of energy efficiency combined with small-scale generation can help meet most of our needs into the foreseeable future, and lessen the need for larger conventional generation. But although we may need more centralized power plants in the years to come, I would not hesitate to rule out big generation projects that depend for their economic viability on power exports and thus require more transmission infrastructure. If we go down that road, we'll be FERCing our future. Transmission superhighways economically favour large centralized power plants, which in North America are primarily nuclear and coal. The nuclear plants will probably disappear in a generation or so, but the coal plants certainly won't, not in the U.S., which has centuries worth of the fuel. If we invest billions of dollars in expensive transmission infrastructure just to give us better access to bigger coal plants, or, even worse, to give Americans access to our always-on coal plants, our children and grandchildren will question our sanity as they cough their way to early graves.

And who will pay for all these shiny new towers? Transmission is not cheap and any investments we make in it are "sunk costs." You can turn an old office building into condominiums or convert an old church to a restaurant. You can even turn an old power plant into a public facility of some kind, which has been done. But a transmission system can be used for one thing only and once it is built it must be paid for whether it is used or not. This was the same conundrum Adam Beck had to deal with when trying to bring the HEPC network to rural Ontario. The lower the demand on the system, the more per kilowatt (horsepower, in his time) the end-use customers had to pay. If we build new transmission infrastructure to facilitate inter-jurisdictional power trading (which is the only reason to do so) we will have to pay for it *forever*, whether or not the new facilities can be economically justified. Estimates of the costs of enhancing the North American grid so that congestion is substantially reduced vary considerably, between $60 and $250 billion. This is over and above normal maintenance costs. Even at the low end, it is preposterous to spend so much money entrenching the most environmentally damaging sources of electricity and discouraging smaller-scale and much cleaner alternatives. I could also talk about the dangers of greater corporate concentration in the power sector, which is the inevitable consequence of a system based on billion-dollar plants. These dangers include greater likelihood of supply manipulation and the immense market clout that comes with such concentration. Even if the few corporations that would control our grid-dominated future were all saintly guardians of the public good, we would still be vulnerable to the cost impacts of unexpected outages of large generating units, as happened in Ontario with the Bruce Power reactor shutdown in the summer of 2002.

So even if the Tower Builders are right, they are wrong. Their prescription for what ails deregulation will have awful and permanent side effects. Theirs is not the future we want for Ontario, or for our world.

18

The Retreat from Privatization

Recent efforts at electric utility privatization in Canada and the United States have fizzled, mainly because of public opposition. The public has been proven right.

Dennis Kucinich is a public power hero. His inspiring story should be known to everyone in Ontario who is opposed to hydro privatization and deregulation. In 1977, at age 31, Kucinich was elected mayor of Cleveland, the youngest-ever mayor of a large American city. He had campaigned on a promise to cancel the sale of Muny Light, the city's municipally owned utility, to its privately owned rival Cleveland Electric Illuminating Company (CEI). The previous mayor, Ralph Perk, had approved the sale. The city desperately needed the money. Perk, who styled himself a fiscal conservative, had racked up over $100 million in debt during his six years in office, despite having sold the sewer system, the sports stadium and the transit system to private interests. On November 8, in what became known as the "Tuesday night massacre," Perk and all seven councillors who had supported the sale of Muny Light, were voted out of office. It was the second election in Cleveland history that had centred on public power. The first was in 1905, when Tom Johnson won election as mayor on the promise to create a competitor to CEI. The Cleveland Municipal Light System was born in 1906, the same year as Ontario Hydro. Over the years, CEI did everything it could to kill off Muny Light, without success. In 1976, however, it got close. Perk said that it was time to get government out of the electricity business. Private enterprise could do it better, even though Muny's rates were as much as 25 percent lower than CEI's. Kucinich, a political activist since his teens, led a community coalition that fought the sale and managed to stall it until he could get elected.

His first act as mayor was to cancel the privatization.

CEI fought back. It needed Muny's 50,000 customers to help pay off the growing debt being generated by its aggressive nuclear construction program. CEI had close business relationships with the bank that held the city's debt. When a $15 million note came due in 1978, the president of the bank visited Kucinich and told him to sell Muny to CEI. If he didn't, the bank would not renew the city's credit. If the young mayor saw reason, however, and agreed to the sale, the bank would give the city an additional $50 million worth of credit. Political blackmail. Kucinich had reduced the city's spending by 10 percent in his first year on the job, but was still paying off bills from the Perk administration. Without access to credit, the city would go into default. Kucinich knew that if he refused to sell Muny Light, he would be ending his political career. He said to the bank president: "Look, I can't do that. The utility belongs to the people, it's not mine to sell."

The bank made good on its threat. Credit was cut, the city defaulted, and a business-led movement to get rid of Kucinich led to a rare mid-term recall election. The city council put the issue of the Muny sale on the same ballot, though, and after a bitter winter door-to-door campaign, Clevelanders voted 2:1 to keep both Kucinich and their public utility. It was the city's third election in which the issue was public power. It was also Kucinich's swan song. The banks continued to refuse to keep the city afloat. Services had to be cut and taxes had to be raised. The Cleveland Plain Dealer editorialized relentlessly against the "so-called Boy Wonder." He lost the 1979 election, although on the strength of the previous year's referendum vote, the new mayor also refused to sell Muny. That year, in the wake of Three Mile Island, CEI cancelled plans for four new nuclear plants after spending hundreds of millions of dollars on early work.

At 33 years old, Dennis Kucinich, who had already been tagged as a future U.S. Presidential hopeful, was out of work and unable to get a job in his hometown. No one would touch the young man who had defied the banks, and lost. He ran out of money, his marriage fell apart and he left Ohio to try to rebuild a life in the West.

A decade later, Kucinich returned to Cleveland at the urging of some old friends and supporters. He was now being belatedly recognized as the one who had saved Muny Light. The utility had not only remained in public hands, but its rates were now even lower, relative to CEI's, than they had been in 1978, by as much as 30 percent. A banner headline in the Plain Dealer summed up public sentiment: "DENNIS WAS RIGHT!" He was honoured by the Cleveland City Council for "having the courage and foresight to refuse to sell the city's municipal electric system." Kucinich was elected in 1994 to the Ohio Senate primarily on the issue of expanding Muny Light, now known as Cleveland Public Power. Two years later, he was elected as a Democrat to the U.S. House of Representatives in a traditionally Republican district and has been re-elected by landslides ever since. In the latest election, he won 77 percent of the vote without spending a dime on television, radio or print advertising. He always campaigns door-to-door and at community meetings. He is a leader of the Congressional Progressive Caucus and is a candidate for the 2004 Democratic presidential nomination.

In an interview 20 years after his decision to save public power rather than his own political skin, Kucinich was asked what he had thought, during his years in exile, of what he had done:

> I didn't realize it then, but I was really being asked to submit to a view of the world that holds that corporate values must triumph over the public good. That's the decision I had to take a stand on, and I tell you, it was a time in America

Dennis Kucinich. Labelled "Boy Wonder" by his admirers and "Dennis the Menace" by his detractors, Kucinich was, at 31, the youngest ever mayor of a major American city. In 1978, he refused to privatize Cleveland's Muny Light in order to save his political career and lost his bid for re-election after the banks, to punish him, denied the city credit that would have prevented default on loans amassed by the previous mayor. Years later, Kucinich was welcomed back to Cleveland as a public power hero and was elected first to the Ohio Senate and then the U.S. House of Representatives, where he is now a leader of the Congressional Progressive Caucus. Muny Light remains publicly owned with rates 30 percent lower than its privately owned rival.

when it was considered unseemly, in poor taste, to even raise the issue. People are now starting to look at the overwhelming influence of corporations in public life and how the public good can be undermined... and there's an increasing awareness of the heavy cost of privatization of public resources.

Public power in the U.S. has become even stronger since Kucinich's fateful decision a quarter century ago. Privatization of public utilities is nowhere on the American political agenda. The trend, in fact, is in the opposite direction. This is especially true since the 1998 "municipalization" of the once privately owned Long Island Lighting Company, a New York utility serving more than one million people. Now called the Long Island Power Authority, the newly public utility reduced rates in one year by an average of 20 percent, the largest across-the-board rate reduction in U.S. history. This was even better than the average public power cost advantage. Residential customers of private utilities in the U.S. pay an average 16 percent more than customers of publicly owned utilities. In Maine, the privately owned utilities are twice as expensive as the publicly owned ones. In only eight states are private power residential rates lower than those offered by public power, and the differences are typically measured in fractions of a penny.

Long Island joined Los Angeles, San Antonio, Seattle, Phoenix, Kansas City, Austin, Memphis, Orlando, Omaha, Sacramento and more than 2,000 other cities in the U.S. that own their own utilities. A few, such as Los Angeles, Sacramento, Phoenix and now Long Island, own generation facilities as well as the local wires systems. Most, however, purchase their power from others and sell it to their customers at cost. Many buy from publicly owned generators, such as the Tennessee Valley Authority (TVA) and Bonneville Power Administration (BPA). Today in America, public power provides 25 percent of all electricity consumed and serves 74 million people, about seven times the population of Ontario. These numbers are expected to keep growing. More than 100 communities are studying municipalization and several have voted for it directly. In the November 2002 U.S. elections, Las Vegas citizens voted to repeat the Long Island experience and buy out their local power company. Even in Las Vegas, where they gamble on just about everything, people aren't prepared to gamble on deregulated, privatized electricity. In many Florida communities, being in favour of private utility municipalization has proven to be a ticket to city council. In Oregon, a powerful public movement has emerged to have the city take over bankrupt Enron-owned Portland General Electric. California is moving its entire transmission system under public ownership; Nebraska's grid has been that way for decades.

In the land that invented the very idea of privately owned utilities, public power is gaining real estate. Why? Because deregulation has fully exposed the stark differences between utilities that are focused on service and those that are focused on profit. For decades, regulation of privately owned utilities had blurred these differences, but no more. Throughout the California power crisis, Los Angeles, because of its publicly owned utility, was an island of price stability and system reliability and remains so to this day. Any proposal to privatize the Los Angeles Department of Water and Power would be either laughed out of town or treated as a script for a horror movie. Nor would the idea fare any better in other American public power cities. Not even Ronald "Government *is* the problem" Reagan ever proposed selling off the federally owned Tennessee Valley Authority or Bonneville Power Administration. Nor would his political grandson, George W. Bush, ever consider this. While U.S. public power is not without its problems, it has proven to be a better way to provide reliable, affordable electricity.

So, if it wasn't in the U.K., the U.S., or, as I have argued in this book, in any objective assessment of Ontario's own history of public power, where *did* the Harris/Eves Conservatives find evidentiary support for their central premise that the private sector can do this whole power thing better?

They certainly didn't find it in the rest of Canada, where utility privatization is also not on the popular political agenda. Quebec, Manitoba, Saskatchewan and British Columbia all have publicly owned utilities and it would be a foolhardy politician who ran on a platform of selling them off. In British Columbia, the Liberal government of Gordon Campbell ran the idea of BC Hydro's privatization up the flagpole after taking office in June 2001, but is now trying to duck and dodge on the issue in the face of stormy public opposition. During the election campaign, Campbell had said he was unalterably opposed to privatization. It seems Liberal policy commitments are as rock-solid in B.C. as they are in Ontario. The Campbell government is now proposing to restructure BC Hydro into separate wires and generation companies, in obeisance to the U.S. Federal Energy Regulatory Commission, but any hint of privatization will continue to be met with great resistance. It's simply not in the cards.

If privatization had a chance of gaining public support any place in Canada, you would think that Alberta would be that place. Think again. The only two attempts in Alberta to privatize municipally owned utilities have been turned back by public campaigns that evoke the early days of Ontario's public power movement.

Edmonton boasts one of Canada's oldest publicly owned electric utilities. The Edmonton Electric Light and Power Company was established in 1891 and became a municipal utility in 1902. One year later, the city's water branch was integrated into the electric utility, a North American first. In 1970, the utility was rechristened Edmonton Power and in 1996, EPCOR was formed as a separate corporate entity with the City of Edmonton as sole shareholder. EPCOR is one of the very few municipal utilities in Canada to own its own generation. In fact, it is the second largest power generator in Alberta, with nearly 20 percent of the province's supply.

In 1998, a move to privatize EPCOR was narrowly defeated by City Council after a citizens' coalition made the proposal a front burner issue and polls revealed sentiment against the sale running nearly five to one. Its continuing public ownership proved no barrier to growth and profitability, however. In 2001, EPCOR bought Ontario's Union Energy and has substantial interest in power projects in British Columbia and Washington State. Seventy percent of its revenues come from outside Edmonton. With 3,500 employees, assets of $4 billion and a customer base of 1.6 million, EPCOR is Canada's largest municipally owned utility. In 2001, the company returned $226 million in profits to its shareholder, the City of Edmonton, about $1,000 per household. If you're an Ontario customer of Union Energy, the people of Edmonton thank you.

Those familiar with Alberta's political landscape might be tempted to say: "Of course, they'd be opposed to privatization in 'Redmonton,' it's the only place where they elect New Democrats in an otherwise solidly Conservative province." But further to the south, in Calgary, where Ralph Klein was mayor before he became premier, privatization of city-owned ENMAX was *the* issue in the 2001 civic elections.

The City of Calgary Electric System was established in 1904 because privately owned utilities had been chronically incapable of meeting the rapidly growing power needs of the bustling young pioneer city. At first the utility generated its own power, but later decided to purchase power from wholesalers and concentrate instead on the wires business. Following Edmonton's lead, the Calgary Electric System was put into a separate entity in 1998—ENMAX Corporation, which serves 411,000 customers in Calgary, Red Deer and Lethbridge, plus a number of smaller communities. In 2001, the utility's reliability performance was ranked among the best in Canada. Its revenues that year were $1.4 billion, with profits of $249.6 million, somewhat less than $1,000 per household. In September 2001, largely driven by ideology, Calgary City

Council put this great public asset on the auction block by voting to seek binding bids for the utility. Kansas City-based Utilicorp, for one, was quite interested, as were ten other companies from around the world. Potential sale prices of between one and two billion dollars were frequently mentioned. Fortunately, the civic election was only one month off, in October.

Public reaction against the proposed sale was intense. Mayoral candidate Bev Longstaff, who as alderman had voted for the sale, claimed privatization would protect Calgarians from the risks of a deregulated market. "The City of Calgary is not losing an asset," she said. "The City would be leveraging an asset, exchanging a risky asset in a volatile market for cash to be reinvested where Calgarians think best." She accused her main opponent, Dave Bronconnier, of fear-mongering for claiming that the sale would not only deprive the city of a steady stream of stable revenues, but also lead to increased power bills. Bronconnier, for his part, promised to cancel the call for bids and begin anew the process of determining if a sale was in the best interests of the city. In an echo of Cleveland's 1977 election, Longstaff was defeated, as were seven other incumbent councillors who had also supported the ENMAX sale. The new mayor made good on his promise and appointed a committee of the council to re-study the issue. On May 10, 2002, Calgary City Council voted 13:2 to keep the utility in public hands.

In Ontario as well, despite the encouragement of the Conservatives, local Hydro privatization has been a non-starter. With intuitive understanding of the importance of democratic control over this most essential of public services, city councils throughout the province have been decidedly reluctant to even consider selling their local distribution systems to private corporations. Those that did investigate it at some point have almost all backed away.

In all of North America, it now appears, the Eves Conservative government, with the support of the Liberals, is the sole political standard bearer of power industry privatization.

In December 2001, the Conservatives announced their intention to sell off Hydro One, which owns the province-wide transmission system, as well as distribution networks that serve over 25 percent of the province's four million retail customers. Mercifully, neither Premier Harris nor his Energy Minister Jim Wilson invoked the spirit of Adam Beck in this truly base and cynical ploy to pay for some pre-election handouts by selling off the Crown jewels. The Initial Public Offering (IPO), the largest in Canadian history, was expected to bring in $5.5 billion that would finance further Conservative tax cuts and other pre-2003 election bribes. But a legal challenge from the Communications, Energy and Paperworkers, and the Canadian Union of Public Employees, threw a wrench into the plan. In mid-April 2002, just days after Ernie Eves was chosen premier as Harris's successor, the Ontario Superior Court ruled that the IPO could not go ahead because the 1998 Energy Competition Act had not explicitly authorized the government to conduct such a sale. To overcome the court decision, Eves introduced Bill 58, to give the government the unambiguous authority to sell off Beck's dream. The new premier publicly mused, however, that he might not actually use that authority, much to the consternation of the Bay Street financial houses. They had been salivating over the prospect of at least $100 million in IPO commissions. To add to the confusion, an unexpected public scandal erupted over the astonishing salary and unusually rich benefits package of Hydro One's CEO, Eleanor Clitheroe, distracting the government from whatever privatization plans it was hatching.

The Conservative government had instructed the management and board at Hydro One to start running the company as if it were a private sector company, so they did. Clitheroe's salary and bonuses were increased from $513,000 a year in 1998 to $2.18 million a year in 2001. All of the other corporate executives at Hydro One received similar absurd pay and bonus increases as well. This is how privatization saves you money on your hydro bill! Clitheroe was eventually fired, as was the entire Hydro One board of directors, but the ensuing public debate over whether or not Ontario would be better served by a privately owned transmission and distribution company forced the government into a temporary retreat.

In July, Finance Minister Janet Ecker revived the public vs. private debate when she announced that the government was seeking outside investors for Hydro One "to upgrade our system while keeping the control of Hydro One in public hands. A strategic partnership will help Ontario to raise needed funds to continue to provide reliable power for today and into the future."

The Finance Minister's bafflegab was preposterous. The government of Ontario does not need help raising money for infrastructure improvements. Lenders are camped outside Hydro One's door, more than happy to provide whatever funds are required for infrastructure, because they know the money they lend has the security of a government guarantee. Ecker wants people to believe that financial institutions will somehow balk at loaning money to a government-owned utility, but will be more generous with a "strategic partner" that has neither regulatory nor taxing authority. This analysis makes no sense, and is therefore right in line with the Conservatives' energy policy.

Months later, the government announced that it would attempt to sell 49 percent of Hydro One, a half-baked privatization scheme that puzzled the credit-rating houses. Standard and Poor's changed its outlook for Hydro One from "stable" to "negative." Other bond rating agencies followed suit. Then, in January 2003, Eves backed away from the idea altogether. Public opposition to even a partial sell-off of Hydro One was so strong he had to try to neutralize it as an election issue. Besides, he couldn't entice corporate interests to invest in something they couldn't control.

The $2 billion or so dollars the Conservatives had hoped to pocket from the sale of our provincial electricity network is a fraction of what they expect to pull in from the planned privatization of two-thirds of our publicly owned generation system. Some of this has already happened. A significant chunk of Ontario Power Generation has already been "decontrolled," a euphemism for privatization invented by some chuckling bureaucrat. Premier Eves keeps insisting that Bruce Power was not privatized, presumably because the province still owns the facility. More bafflegab. Look the people of Ontario in the eye, Ernie, and tell them the truth: "A private company, Bruce Power, has a long-term lease to operate the Bruce nuclear station. It charges us what it can get for the power it generates and gets to keep all the profits, which have been considerable. But fear not, we still technically "own" the facility, which means we are responsible for the perpetual nuclear waste and costly decommissioning obligations after Bruce Power has made off with the profits." Then explain how this is a better arrangement for Ontario taxpayers and consumers, now and in the future. Don't forget the part about the bankruptcy of the privatized British Energy, original majority owner of Bruce Power.

Not as large as the nuclear station giveaway but even sadder, was the March 2002 sale of four hydroelectric generating stations on the Mississagi River, east of Sault Ste. Marie, to a group led by international resource conglomerate Brascan. (Ironically, Brascan was founded by Syndicate partner William Mackenzie, whose private power interests in Ontario were bought out by the provincial government in 1920.) The stations were built by Ontario Hydro between 35 and 55 years ago and together produce about as much power as Sudbury—a city region of 200,000 people—uses. Sale price of the power dams: $340 million, a small fraction of the total cost of the Conservatives' hydro rebate scheme. During their first summer of private sector operation, Brascan completely drained the power dam's reservoir, Rocky Island Lake, killing uncounted numbers of fish and sending an ecological shock wave up the environmental chain. To hell with Mother Nature. With Ontario spot market power prices in the summer averaging *nine times* the cost of producing power from these stations, the fish never had a prayer. Ontario Hydro, by contrast, had never depleted the reservoir, even in relatively dry years. In fact, Hydro regularly consulted with organizations representing anglers, cottagers, commercial resort owners, First Nations communities and other water use stakeholders all over the province. The idea was to ensure that public water resources were equitably shared. In Ernie Eves's Ontario, equity is for wimps, profit is for Ernie's friends, like Brascan, a huge contributor to his leadership campaign and the Conservative Party.

An NDP government would take back these stations and refund Brascan's money. We agree with that former Conservative premier who hated the idea of the province's hydropower being made the sport and prey of profit takers. As for the Bruce Power lease, we would let that run its course in a re-regulated marketplace. We would rather spend public money on reducing our dependence on nuclear stations, not supporting it. We would rely, as we must, on the Canadian Nuclear Safety Commission to ensure that the station is operated safely.

Asserting that something is true does not make it true. The claim of the Liberals and Conservatives that an unfettered private sector is inherently better than the public sector at meeting our society's power needs is an assertion that miserably fails the test of experience. There is simply no credible evidence whatsoever that private power is more efficient than public power. Hearing Conservative front benchers bleat about the need for market discipline in the energy sector makes me want to gag. For Enron and others of that ilk, "market discipline" is an oxymoron. These people had no discipline. They cheated, lied and stole, and then tried to cover it up by cooking the books. Though crooks like this are always in the minority, our electricity is too important to expose its supply and price to their greed and dishonesty. And even if we could concoct a perfect set of rules that would prevent their market manipulations, even if we could actually build that Holy Grail of the marketplace—the perfectly level playing field—it is highly unlikely that private power would prove better than public power. This has nothing to do with ideology, but with the inescapable economic fact that most of the costs of power systems are tied up in capital and fuel. Finance Minister Ecker's delusions about needing a "private partner" to get financing notwithstanding, government has the clear advantage in capital costs, and government utilities have no disadvantage in fuel costs. The rest is essentially labour costs, which represent a small portion of the power bill. If you were a true believer in deregulation, you would assert that the private sector is better at managing all these costs and you would grasp for anecdotes, such as Darlington, to substantiate this. Let's trade anecdotes:

- How does the fact that Manitoba Hydro has the lowest power prices in the Western hemisphere, and possibly the world, square with your notion that the public sector is worse at managing costs?
- How do you explain that in the United States private power is, as we have seen, 16 percent more expensive than public power? Why is it that Cleveland's publicly owned utility sells power for so much less than its privately owned one?
- Why haven't public power customers in Los Angeles seen a price hike in ten years?
- Why did pre-deregulation Ontario Hydro, despite its faults, have significantly lower unit energy costs than those of every privately owned utility in nearby U.S. states—in almost every state, in fact?
- How is it that the generating stations of publicly owned EPCOR compete so well against Alberta's privately owned generators?
- Why have the credit ratings of so many privately owned U.S. utilities plunged since 2001, with several once-mighty companies like Virginia's AES Corp., which owns an interest in 177 generating assets in 33 countries, now reduced to the status of junk bond issuers?

Okay, your turn.

The non-ideological advantages of public power go beyond price. I have talked a lot about energy efficiency and it is self-evident that profit-driven utilities in an unregulated marketplace have absolutely no economic interest in lowering demand. Their interests obviously lie in creating more demand and it is unreasonable to expect otherwise. Our society, though, has a very important interest in greater energy productivity and its resulting economic and environmental benefits. A public power system does not have the inherent conflict of interest between profitability and environmental sustainability that structurally burdens non-renewable private power. I applaud those private corporations that do their best to reduce their environmental impacts, but I know from my time as Minister of Natural Resources and as Attorney

General that we would be foolish to put industry in general on the kind of environmental "honour system" that characterizes the Conservative approach. The fact that Conservative and Liberal governments have done such a bad job at promoting energy efficiency and environmental protection in the past does not mean that all governments are incapable of doing so. In every province where it has held power, the NDP has proven otherwise, as have many other governments around the world.

Economic development is another critical area in which public power is clearly the better way. It always has been. When Ontario's public power movement began one hundred years ago, it wasn't driven by an ideology that public ownership was an inherently superior economic model. Snider, Detweiler, Beck and the other leaders of the movement were not socialists. They were municipal leaders and business owners who were focused on economic development for their own communities and, of course, for the province as a whole. They represented residents and manufacturers who were tired of paying high prices for unreliable power. They were worried about Ontario's heavy dependence on imported energy, which at that time was coal brought up from the United States and burned here to produce electricity or steam. Today, the forces of deregulation and private power have brought us full circle, back to high prices and a reliance on imported energy, which in our time is still coal, but which has already been converted to electricity before being shipped here.

Since I am so critical of free market ideology, at least when it comes to electricity (and education and health as well, but those are other books), I do not want to be ideological myself and assert that privatization in the power industry is always going to turn out for the worse. If we look hard enough, we may find privatization scoring a few short-term triumphs. Where an industry's technology has been changing rapidly, for example, private interests can bring a breath of fresh air to enterprises that have grown a bit sclerotic. But this argument applies indifferently to public and private enterprise. Thus, Bill Gates and Microsoft stole a march, not on any government bureaucracy, but on the mighty and thoroughly private enterprise, IBM. And for a brief period of time, because it was more in tune with new technology, Nortel was able to outflank AT&T, at a time when AT&T still owned Western Electric and was in the telecommunications equipment business. What we are dealing with is not the issue of public vs. private, but of the established vs. the innovative.

The system most geared to getting fresh blood into any system, whether that "system" is commercial, governmental, social or even familial, is democracy. That is why New Democrats, like me, place so much emphasis on democratic control. I firmly believe that to the extent that government was not as fully responsive as it should have been in the past, in relation to the management of public power or any other resources which belong to all of the people, it is because government has forgotten its basic mandate to be eternally responsive to the needs of its principal constituency, the public as a whole. Pay attention to public need and not private greed, and you will have the basic formula for making sure that government itself and the essential services we want our government to provide are both efficient and innovative.

19
A 21st Century Public Power System

Ontario's future power system must serve its economic and environmental needs. Public power is the only way we can ensure democratic control over our energy future.

What if we could start all over again? How would we design Ontario's power system today from the ground up? What have we learned over the last century and how would we best apply that knowledge if we had to build a system for the next hundred years? This is not an idle thought experiment. The Conservatives, abetted by the Liberals, already *are* designing the province's electricity system for the twenty-first century. Fundamental decisions about deregulation and privatization have been made without any consultation or discussion with the public. Thankfully, these decisions are not irrevocable. We can take a different path. We can certainly cancel deregulation, as California has done and others are seriously considering. We can reverse what privatization has happened. There is little point in doing this, however, unless we know why we want to do it.

Let's start with electricity system principles that would be readily agreed upon by a very large majority of Ontarians today.

1. We need **Reliability**. This is the number one requirement of a power system. We have to know that the power will always be there when we need it. We don't want to live in suspense every time we reach for the light switch, as Conservative Energy Minister John Baird so cavalierly suggested in early December 2002, when Ontario came close to blackouts.

2. We want **Stable and Fair Prices**. Stability is key. We don't want

the opening of the monthly hydro bill to be a dramatic ritual. Basic economic security requires that we have a general idea of what our hydro is going to cost from month to month. At the very least, it helps with budget planning. It also allows us to better calculate the economic benefits of energy efficiency and conservation measures.

By fair prices, I mean prices that are based on the actual cost of producing the power. And this cost must include, insofar as it is possible to calculate them, the external costs of production, such as the impacts of fossil fuel pollution and the still unknown price tag for permanent disposal of nuclear wastes. Some of these indirect costs may be difficult to quantify with precision, but we are deluding ourselves by pretending they don't exist, unfairly passing them on to future generations. We must make our best efforts to incorporate these "externalities" into our regulated prices. You will notice that I did not include "low prices" as a defining principle of our future power system. "Low" is a vague and highly relative term when it comes to power pricing and can be misleading if it does not include externalities. Fair is better than low.

3. We must promote **Energy Efficiency**, for both economic and environmental reasons. If we could get cold beer and hot showers using half the energy we now use, who wouldn't want to do this? Mother Nature will be cheering on our efforts.

4. We want **Environmental Sustainability**, which means steadily reducing the environmental impacts of our energy use and moving ever closer to the ideal of a pollution-free and completely renewable energy future. In the words of the 1994 Ontario Hydro Annual Report: "A global consciousness that all nations and cultures share a common environmental future has changed the world's view of economic progress. There is now widespread agreement that present generations must endeavour to meet their own needs without compromising the ability of future generations to meet theirs." If you do not personally share this view, or something like it, I am surprised that you made it this far into the book.

5. We need to be able to use our power system for **Economic Development**. Indeed, a reliable supply of electricity at a fair

and stable price is a fundamental underpinning of a modern economy. This was the original purpose of Ontario Hydro and I think we can all agree that it has fulfilled that purpose admirably. But what about the future? Will there be opportunities to use our system for further economic development? To the extent that this is possible, I am quite sure most Ontarians would support it.

6. Overarching all of these principles is **Democracy**. We want someone who is publicly accountable to the electorate to take responsibility for the future of our electricity system. Even those who do not believe in public ownership, like the Liberals, would not let government off the hook if our power system deteriorated, reliability suffered, prices skyrocketed, nuclear wastes were improperly tended and fossil emissions from coal generating stations coated their children's lungs. *Someone* has to take responsibility for our electrical future; it is fundamental to social and economic order.

These are the six principles that I firmly believe most people would want to see embodied in a future electricity system. Reliability. Stable and Fair Prices. Energy Efficiency. Environmental Sustainability. Economic Development. Democracy. Now let's analyze whether these would be better served by a regulated public system or a deregulated private system.

Reliability has two aspects: supply and delivery, or generation and wires. Is there enough power available and can it be carried to where it is needed? Let's consider the delivery system first—the transmission and distribution wires networks. In this aspect of reliability, a regulated, publicly owned system wins hands down. Even Liberals and Conservatives agree that wires networks should be regulated to ensure that they are properly maintained and technologically upgraded; they just think that a private company would do this better than a public one, or more cheaply. There is not a shred of evidence for their belief. Ontario's transmission system, to pick the example that concerns us most, has always been one of the sturdiest and most reliable power networks in the world. It is also one of the continent's largest, at 30,000 km, and traverses a number of challenging climatic zones. The workers who built it obviously knew what they were doing. Those who maintain and repair it are second to none in both skill and dedication. They can even service high voltage transmission lines that are still "live," a remarkable innovation Ontario Hydro helped pioneer some years ago and that has been adopted in many other countries. Indeed, publicly owned Ontario Hydro was from its very beginnings a technological leader in the transmission sector. Hydro's own High Current Test Facilities are recognized as a major research and equipment certification centre. All this has been achieved quite efficiently and economically, since the time Adam Beck's young engineers delivered the first line from Niagara to Toronto for 25 percent of what the private sector said it would cost.

Are engineers in publicly owned utilities today somehow not as talented or hard-working as those in privately owned ones? I very much doubt that.

On the front lines of the system, much of the efficiency and reliability of Ontario's grid is attributable to a highly experienced, well-trained, well-paid maintenance workforce. Anti-public power types would claim that a private company could drive down transmission system labour costs without driving a skilled and experienced workforce away. There is no evidence to support this assertion. (In fact the recent experience with nurses and other health care providers in Ontario shows that such a strategy drove nurses out of the profession or out of the province.) Nor do I think it wise to introduce Mr. Profit to Ms. Maintenance, certainly not when it comes to wires networks. It would not be a happy marriage. The two have little in common. Yes, you can regulate performance measures that supposedly keep for-profit wires company operators from unduly skimping on maintenance, but let's be real about this. When the pressure is on to keep costs low and margins high, something has to give. In large, complex systems like the transmission grid, there would always be hundreds of opportunities to shave maintenance quality, and thereby costs, without short-term consequences. But what about five or ten years later? If I am the CEO of a private transmission company holding stock options I hope to cash in, am I going to be eager to spend the money to ensure a grid that will be solid for another 30 years, even at the expense of immediate profits, share prices and my own annual bonus? There would be a fundamental conflict of interest between *me* getting rich quickly and the long-term reliability of *your* power system. Ontario has had a well-run wires system. Why take the chance of turning an essential service that we need every day—the electricity transmission grid—over to an Enron or British Energy, a WorldCom or a Nortel?

All of the above observations apply equally to distribution networks—the municipal Hydro companies and Hydro One Retail networks. They are efficient and well-run now by public employees. What would we gain from privatizing them? Only the obligation to line private shareholders' pockets every month, an obligation that would cost us more money on the monthly hydro bill and quite likely lead to less service and less reliability.

In summary, if we want our future electricity delivery system to maintain the sterling reliability we have come to expect in Ontario, regulated publicly owned wires companies and local hydro commissions are the way to go. That is why, on December 13, 2002, I publicly announced that an NDP government would buy back any portion of Hydro One sold by the Conservatives. I said we would simply refund the purchaser's money. My objective was to serve notice to all potential investors that they should back off any deal until at least after the next election. Five weeks later, Ernie Eves announced that the sale was off. This was a partial victory but it won't be complete until the legislation allowing the sale is repealed. The people of Ontario voted to create a public power system—in two referendum votes, remember—and the NDP will oppose any attempt to sell any part of it off without another vote of the people on that specific issue.

On the supply side—the generation of electricity—it has already been well-established through recent experience that deregulation is bad for reliability. In a profit-driven market, the less power available, the higher the price can be driven up and the more money the available suppliers make. This inverse relationship also works the other way: the more power available, the less money generators make. There is simply no way of resolving this inherent conflict of interest in deregulated markets except through elaborate and fragile market mechanisms, eternal vigilance by system managers and, ultimately, some hybrid form of regulation and bribery (to induce private companies to build more supply). It is quite clear that regulated electricity rates are far better for ensuring reliability. Think of California, where the power shortages ceased immediately after regulators moved in with price caps. For reliability reasons alone, regulation must continue to be a central feature of Ontario's hydro system.

Stable and fair prices can only happen through regulation, for reasons already fully explored. Of course you *can* get medium-term stable prices in a private power market by signing one of those fixed-rate contracts offered by deregulated retailers. Last summer, however, hydro consumers learned the hard way that signing on with an electricity retailer for three to five years meant paying 40 percent more for power. They got a stable price, but was it fair? If you believe it's fair to charge a 40 percent premium for price stability, then why did Ernie Eves implement a price of 4.3 ¢/kWh in November when the retail marketers were charging 5.9 cents/kWh? The answer is because the public did not think it fair that prices should be so much higher than they were only a few months earlier. Too bad the premier isn't also prepared to listen to the public on the issue of Hydro privatization. Seventy percent oppose it.

Energy efficiency, as we know, is of no interest to private power producers in a deregulated market. They want to sell more, not less, so they can make more money. There is, nonetheless, a role the private sector can play in our efforts to reduce energy use. Demand-side energy efficiency is local by nature. It happens building by building. There are millions of buildings in Ontario; few of them are anywhere near as energy efficient as they could be. Using currently available materials and technologies, Ontario could reduce its energy use by at least 40 percent in the residential, institutional and commercial sectors. In virtually every region in the province there are engineering firms and local contractors who could be put to work making homes, offices, stores, schools, hospitals and factories more energy efficient. Moreover, Ontario industries that manufacture energy efficiency products would prosper. Good jobs would be created everywhere in the province. Homes would increase in value. The environment would benefit greatly. Far fewer generating stations would have to be built. And all of this would be quite affordable.

Up-front financing of energy efficiency measures is required, just as it is when you're building a generating station. The funding for energy efficiency measures in your home or business could be handled through a non-profit public agency. The NDP would create one called Efficiency Ontario. Its job would be to set and enforce building retrofit standards, recommend the best technologies and practices, certify energy efficiency contractors, monitor results and educate consumers about conservation and efficiency. Working with local hydro commissions, Efficiency Ontario would advance the money to pay for measures that would *permanently* reduce your building's energy use, both gas and electricity. It would then recover this money over time on your hydro or gas bill. The rate of recovery would be set so that your post-retrofit bill is no larger than your pre-retrofit bill, making the whole thing effectively free to you. Your monthly repayments would only be the difference between your old hydro or gas bill and your new one. Once your retrofit costs were repaid, you would keep 100 percent of the savings for as long as you occupied the building—your energy efficiency dividend.

The initial seed money for Efficiency Ontario would come from a small charge, a tiny fraction of a cent per kilowatt-hour, on all electricity sales in Ontario. This is not a new idea. Fifteen U.S. states have similar "system benefits charges:" Ohio, California, Illinois, Massachusetts and New York among them. Even Texas does this. Most other states are seriously studying the idea. The money taken in from Ontario's energy efficiency charge would not be used to directly finance retrofit measures but would provide "insurance" that would attract large institutional investors with patient money, such as pension plans, into long-term funding of energy efficiency. Efficiency Ontario would be an example of a public sector strategy that creates productive opportunities for private sector firms. Each sector would do what it does best and local economic development would be the result. While in government, the NDP took this approach with our Green Communities program. It was warmly received in many communities and did some excellent work. The Conservatives killed it immediately; profit-driven power companies don't want to see effective energy efficiency strategies implemented on a broad scale with public support.

When it comes to environmental sustainability, we cannot rely on publicly owned utilities being voluntarily better in this regard than privately owned ones. Strict environmental regulation by government is, without a doubt, required in either case. With publicly owned utilities there is more political accountability, but if you have an environmentally insensitive government like the Harris/Eves Conservatives, that accountability may not be worth much. Even with strict environmental regulation, a deregulated energy marketplace is bad for the environment. Using Ontario as an example, here's why:

The NDP has long called for conversion of Ontario's coal-burning plants to natural gas. There would be a huge environmental benefit to this, so huge that even Conservatives and Liberals now support gas conversion. But there is a cost to this. Power from natural gas stations can be more expensive than coal, depending on the cost of the gas. Without environmental cost accounting, we continue to give dirty coal a marketplace price advantage. In the Ontario marketplace, the Independent Market Operator is *required* to accept the cheapest bids from generators wanting to sell into the spot market. So the deregulated market that has been established is itself a barrier to environmental-friendly generation that may cost more than conventional power. I know the open marketplace also allows customers to contract for green power, paying more if they wish. Good for those who do that. But this is hardly a visionary approach to the promotion of low impact and renewable energy.

What Ontario needs is a political decision that the province move steadily towards renewable energy, by law. The mechanism is called "Renewable Portfolio Standards," which means that a specified percentage of generation must come from renewable sources. Several European countries and some U.S. states are doing exactly this. The United Kingdom, Germany, Denmark and the Netherlands, for example, are mandating massive wind developments, most of them offshore, since wind tends to blow more steadily over water. We can and should do the same thing here in Ontario. And not just with wind. There are thousands of undeveloped run-of-the-river small hydro sites throughout this vast, water-filled province. Ontario Hydro was guilty of turning its nose up at such small-scale projects, which is why few were ever developed. An NDP government would change that and require all generators to begin producing power right away from renewable sources. By 2020, at least 20 percent of all power in the province would be renewable. This would not be a stretch; it is very achievable. Denmark, with half the population of Ontario, already produces nearly 20 percent of its power from wind alone. Surely we could do this ourselves over the next 17 years, in addition to tapping into our abundant small hydro and biomass energy potential. The $2.5 billion (and counting) that is being spent by the Eves government on the refurbishment of Pickering A's 2,000 nuclear megawatts would have been much better spent on renewables. Do the math, Ernie.

Through regulation, we would blend in the initially higher cost of renewables with the cost of cheaper power and give *preference* to power from clean sources, relegating our fossil fuel further back in the pipeline. As time and technology advances, the cost of renewable generation will come down; this has been happening steadily for the past 20 years. Manitoba Hydro and Hydro Quebec, both publicly owned, not-for-profit power companies, are actually doing this now with wind turbines. Most of Ontario's future renewable generation will be publicly owned because the borrowing and financing costs for the public sector are, as always, so much lower. But some will be the product of community non-profit co-ops, and some could be privately owned by companies that agree to work within the regulated, not-for-profit power system.

We know economic development would be a natural consequence of a prolonged, province-wide push on energy efficiency. In the past, using Hydro for economic development has usually meant subsidizing rates for industries to induce them to locate or expand here. Industrial energy price subsidization can be attractive in theory, but tricky in practice. How can we offer low hydro rates to one auto plant and not to all? If Ontario customers want to subsidize industry generally, that is a political decision that should be made openly. I think it far better to work with industry to lower its energy costs through greater efficiency, not through a scheme of subsidized rates. Whatever the decision, you can do so much more through a regulated public system than you can as a handcuffed observer of market forces.

Democracy must be paramount when it comes to fundamental decisions about the future of our power system. Ontario invented this idea and it is a great one. We are foolish if we abandon it. How do we embed political accountability in our public power system without turning it into a political football, as Mitch Hepburn and Mike Harris did, and Ernie Eves is now doing as well? There is no simple answer to this because the economic, environmental and social issues entangled in those wires we hardly notice running along our streets are far too complex to be reduced to any formula you or I could concoct to resolve them. We must depend on democratically based processes and clear accountability for decisions arising out of those processes. My approach would be to create a Public Utilities Commission with the authority, expertise and resources needed to direct the province's power system in accordance with laws passed by the Ontario Legislature. The existing Ontario Energy Board is supposed to do this job, but the OEB is, in my view, seriously underfunded and lacks the expertise for the critical tasks it has been given. The best electricity system expertise available should be available to the Public Utilities Commission. Other relevant expertise must be available as well, which is why I believe the Commission should have funding for consumer and environmental research, as well as for experts in these areas to testify in open hearings. Does anything less make sense if our objective is to shape the future of our power system in accordance with the principles we have discussed in detail thus far?

There are some important transitional issues that will have to be addressed as we move towards our renewed public power system. This includes the old Ontario Hydro debt, which has been made worse by the

Conservatives by nearly one billion dollars a year since taking office. In 1995, when the NDP left office, there was a sustainable plan to pay down the debt. It *was* working, as Ontario Hydro's financial reports for 1994, 1995 and 1996 show. Political meddling has since driven it up. We have to deal with that. We must also re-examine the way the Conservatives divided up Hydro's debt. Our goal must be to ensure that all customers, from the largest industries to the smallest household, pay their fair share over time, but no one sector is unfairly burdened.

Even as the Conservatives bungle the Hydro debt repayment schedule, they wave their arms hysterically whenever the subject is raised, blaming everybody in sight except themselves, the party that brought you a nuclear Ontario. Fortunately, the Hydro debt, though much larger than it should be, is not the economically crippling burden they portray. Hydro Quebec's debt is much larger, at over $41 billion. Quebec's population is significantly less than Ontario's, so its per capita debt is much higher. Is anyone predicting fiscal doom for Hydro Quebec? Not the international bond rating agencies. It would be my preference, and that of most Ontarians, I believe, to pay down the old Hydro debt sooner rather than later. But first we must get it under control, as the NDP government did beginning in 1993.

There are other transitional issues, besides the debt, as we move from the chaotic mess the Conservatives have created to the kind of electricity future most Ontarians want:

- We should gradually return our local distribution companies, the local hydros, to non-profit status, the way they were before deregulation. This would allow for incremental local rate reductions and needed system investments. Today, a lot of municipalities are using the money they make from their municipal hydro utility to cover the cost of all the services the Conservatives have downloaded onto them—water, sewer, public transit, housing. Funding for these important services should come from our federal, provincial and municipal taxes, not from rate increases or additional charges on our hydro bills. It's wrong, and an NDP government would stop it.

- No more nuclear stations should ever be built in Ontario. Do we need to learn this expensive lesson again?

- Rates must be set "at cost." As you can no doubt appreciate after having read this book, determining the actual cost of our hydro is no simple matter. But we did it in the past and we can do it in the

Construction of the Toronto Renewable Energy Cooperative's landmark wind turbine on the Canadian National Exhibition grounds was financed through the sale of $500 blocks of dividend-paying shares to 460 people. The clean power it generates is sold through Toronto Hydro. Commissioned in early 2003, the "Windshare" project is a pioneering example of the environmentally friendly role that private funding can play in a public power system. *WindShare*

future. A well-resourced Public Utilities Commission can be trusted with this task. One thing is clear, we do not want to jack up the Hydro debt by paying less than the true cost of our power today. That would not be a "fair" rate. Eves and McGuinty have made much of their temporary 4.3¢ rate cap. If you really believe the Ernie Eves promise of "no rate increase until 2006 in a still-deregulated wholesale market," I have a used nuclear plant I'd like to sell you. The reality is if either the Conservatives or Liberals form the next government, they will invent some excuse to kill the rate cap within months and entrench hydro deregulation and privatization. Our hydro bills will be higher than ever.

• We should explore what it would cost to enter into long-term contracts with Manitoba and Quebec for their surplus hydro-electricity. The mega-projects of Hydro Quebec, in particular, are certainly not as environmentally benign as smaller hydro facilities, but the hydro dams are already in place and we in Ontario want to reduce our dependence on fossil and nuclear as we move towards a fully sustainable and smaller-scale power system.

One last question remains: what is the role of private power generation?

There are some exciting new technologies that in the near future will allow those homeowners, small businesses and farmers who want to generate their own power to do so affordably. The day is not too distant when many homes or small businesses could house a hydrogen fuel cell that generates electricity for that building alone, much like the furnace in the basement generates heat. Such fuel cells already exist and are being tested, but they are not yet commercially competitive. Some day they will be. Even major oil companies, like Shell, are investing heavily in hydrogen fuel cells because, as the company says: "We want to be selling energy in 50 years when there's no more oil."

Photovoltaic (PV) cells that generate electricity from the sun can be put on your roof, like shingles. It's an expensive way to produce power now because the capital costs are high. But this will not always be the case. "Thin film" PV technology is expected to be competitive with conventional generation available in ten years, not a long time when we're thinking decades into the future. Small wind farms can be customer-financed, as the Toronto Renewable Energy Co-operative has done with "Windshare," a pioneering wind turbine project now generating power at Exhibition Place. Agricultural operations are a natural venue for biomass generation, converting farm wastes to energy—it's already happening commercially. Geothermal heat pumps are gaining ground, especially when installed in small newly constructed buildings. The French are even working on a wall paint that would convert heat and light to electricity.

What should be the response of a publicly owned, not-for-profit power system to these new technologies? If you want to generate your own power at home or your business or through a local power cooperative, and you can do so in a safe and environmentally responsible way, good for you. We are not interested in owning your fuel cell or solar cell any more than we want to own your furnace. We don't want to own the paint on your wall. Indeed, the public power system should *help you* own your fuel cell or rooftop solar panels or your share in a wind farm if that makes economic and environmental sense for society as a whole, which I believe it does.

Will all this on-site generation technology take us to a completely wireless world? Eventually perhaps, but not for quite a while. Even if you want to produce homemade power, you will still need to be connected to the wires network in case your own equipment fails or requires servicing, as will surely be the case sooner or later. You should pay for that backup power insurance through a fair and reasonable wires connection fee, so that the public network can be well-maintained and technologically up to date. Any surplus power you generate can be fed back into the network.

In some places today—California is one—you can legally make your meter run backwards by feeding energy generated by your solar cells, for example, into the distribution system. The same "reverse metering" principle could apply to any on-site generation. In Toronto, a related concept called the Virtual Power Plant is now under development. Pioneered by long-time energy efficiency advocate Alex MacDonald, the VPP would use the billing system of the local utility to aggregate the electrical output of hundreds of tiny, high-efficiency generators located in commercial buildings. The power would be sold into the local distribution market. The result of the aggregation will be lower financing costs for natural gas micro-cogeneration, fuel cells and solar electric generators, thereby reducing the cost of such environmentally benign power. This same idea could be cloned in every urban centre. In rural areas, agricultural wind farm cooperatives could harvest the breeze and sell it as energy through the public power system. A regulated public power system would be the best mechanism to determine fair prices for such privately generated power, as well as the rules under which it could be fed into the system.

And when the wind blows and the water still flows during the night, when demand is low and there is no need for the surplus power, the energy produced while we sleep could still be economically used to break down water by electrolysis into its constituent atoms —hydrogen and oxygen—with the hydrogen being stored for later use to run fuel cells. Indeed, all forms of renewable energy could, even now, be used to turn ordinary water into potential energy in this way. The U.S. futurist Jeremy Rifkin has recently predicted that the combination of renewable energy generation and hydrogen fuel cells will completely transform the world's energy future within the lifetime of many people reading this book. Using renewable generation to create hydrogen fuel would, in effect, conquer the problem of how to "store" electricity.

Realistically, we are many decades away from the hydrogen economy envisioned by Rifkin and others. It will be a long time before all the power we need can be generated locally and we can take down those transmission towers. I do believe that this is the vision we must work towards and I am certain a public power system would bring this to pass much more quickly and economically than a private power system. Until that day comes, we will continue to need central power plants that feed the transmission system that connects us all. So what about private ownership of these generators? Why do they have to be publicly owned if access to the public network and the rates that are charged are both regulated? Why not, as many have argued, share the capital and operational risks of generation with private investors, especially if they're willing to invest in environmentally friendly generation? After all, a sizeable wind farm could be a central power plant too. As long as we control the terms and conditions of network access, where's the public downside?

My answer to this is not ideological, but economic and environmental. The argument that as long as you regulate access and rates, ownership is indifferent, is once again okay in theory. You have to *assume* that the privately run generators are going to be built and operated more efficiently than publicly run ones, at least enough to offset the 15 percent or more profits that the private plant must earn. In the real world, privately built generating stations are more expensive because the interest rates they must pay on construction financing are always higher than government borrowing costs. And even if we make the unproven assumption of greater private sector efficiency, there would still be solid practical reasons for maintaining a very high degree of public ownership of Ontario's centralized generation, especially in a system that is in technological transition and is focused on energy efficiency.

Public power in Ontario was founded on the economic truth that government-backed utilities—in a stable democratic country—can get significantly better financing terms than those that are not. This will never change. If you're looking for lenders to take a risk in your power generating station and you're not a public entity, the next best thing is to have an ironclad *contract* with a public entity. Then you'll get closer to the financing rates extended to government. Not as low, but closer. Without that ironclad contract, however, forget it. You're too big a risk. You'll have to pay a premium to your lenders, if you can find any. This means that in order to obtain competitive financing, private generators will have to have long-term contracts that lock in a public commitment to buy power from them. But if we lock in too much private power, our system would become more resistant to technological and environmental advancement.

Let's say, for example, that we contract for 50 percent of our presumed power needs, for 20 years, with privately owned natural-gas powered stations. What we have done is lock up a lot of money for a long time that cannot be used on energy-related economic development, energy efficiency or on new generation technologies that are more environmentally benign. Think about the implications of this. If our energy efficiency efforts are highly successful, if there is a technological leap in environmentally friendly self-generation, or if there is a recession that reduces power demand, as happened in the early 1990s, the private generating stations would *still* get paid, by contract. The publicly owned generating stations that provide the other 50 percent of our power would have to take the entire economic hit of overall reduced demand. This makes no sense, economically or environmentally.

Publicly owned generation is ultimately more flexible because we can, for example, make politically accountable decisions to retire fossil fuel or nuclear plants before their useful economic life is over. We may well want to do this in order to have cleaner air or to accelerate energy efficiency and renewable energy instead of pumping more money into keeping these old dinosaurs alive. Such an early retirement of an old fossil station may have a cost, but whatever it is, a regulated public system can ensure that we share that cost equitably, just as we equitably share the cleaner air that would result. Even more important, in my view, is that we can democratically decide that we are willing to pay that cost for clean air.

A health care analogy will illustrate my point. Let's say we had a long-term contract with a privately run chain of MRI diagnostic clinics that used machines that were state-of-the-art at the time the contract was signed. Along comes a technological breakthrough that makes the old MRI machines either obsolete, more expensive or less useful for diagnostic purposes. It could easily happen. Medical advances are announced daily. If these old MRI machines were publicly owned, we could publicly decide to cut our losses and go with the new technology. If we're contractually locked in to the old MRI technology (and the infrastructure that supports it), our economic losses will be greater and our health care standards will fall behind.

There are limits to this argument, I acknowledge. I believe, as a general rule, it is best that economic enterprise risk be distributed as widely as possible throughout our society. I believe in employee ownership, for example. Workers should have a stake in the growth and success of the companies for which they work. In the special case of our power system, however, a very large part of the economic risk is inescapably borne by the public, even if generation is privately owned, simply because we cannot function as a society without electricity.

Our bottom line is this: if we want to ensure that our power system in the twenty-first century operates—and evolves —in accordance with the six principles that the vast majority of Ontarians would support, public power is the only way to go.

20
Our Gift to the Future

The vision of our public power pioneers helped make Ontario one of the world's most prosperous regions in the twentieth century. It is our turn to repay the favour for generations to come.

Human societies need goals every bit as much as human beings. Individually and collectively, people are driven by dreams of a better tomorrow. The need to believe that the future can and will be better is deeply embedded in the human psyche. A continuously renewable reservoir of optimism gets us out of bed in the morning and keeps us going through our often complex and difficult days. We discover anew how important optimism is in our lives every time we lack it, at those moments when we feel there is no point to what we do, nothing worth working for or saving for. When we stop caring about the future, we stop caring about life. Most of us go through moments like this from time to time, bereft of optimism, feeling empty inside. But then that mysterious reservoir of hope starts to fill once more, and sooner than we could have guessed, we are thinking and acting again as if the future mattered. This instinctual conviction that we *can* and *should* make the future better is the foundation of all human progress.

The future we care about is both personal and social. We want prosperity and security for ourselves, of course, but we also understand, each in our own way, that we are more likely to achieve this in a prosperous and secure society. Even if we did not see our own personal future tied up with that of our society, we cannot ignore the reality of this interdependence in the case of our children and grandchildren. I have no doubt that the phrase: "We want our children to grow up in a better world," is a common one in every language. The fact that this is not simply a narrowly focused genetic drive for our own offspring to enjoy a better future is proven every day by the many childless people who devote much of their time to selfless purposes that will outlive them. Their hopes for a better world are no less ardent than those of new parents gazing lovingly on their peacefully sleeping baby. Nor does age wither our intrinsic belief that we can, if we have the vision and the will, make life better for future generations and give them the opportunity to do likewise. Our experience makes us understand that progress is more complex than we may have once thought, but does not dissuade us from trying.

So it was with the people who conceived and built Ontario's public power system. Many of the faces that stare out at us from old photos of hydro-related events are well-lined and weathered. But everyone in those pictures, young and old, knew that people yet unborn would benefit from their efforts even more than they would themselves. Their dream was not just to make their own lives better, but to make it better for those to come. For us. They knew the public power system they were building would long outlast them and *that* was the most exciting part of the dream. They had a vision of a province in which "every small village" could support manufacturing and "the poorest workingman will have electricity in his home." They saw the leading role that hydropower, that wondrous gift of nature, would play in achieving those goals. I am inspired whenever I think of how much we owe them.

It is now our turn to envision anew the shape of Ontario's energy future. One hundred years after the birth of Ontario's public power system, we have exactly the same decision to make as our visionary forebears. We already know the Conservative and Liberal agenda. Both voted to abandon our public power system and deliver our future largely into the hands of private interests. Their vision of the future is narrow and mean: a daily (hourly, in fact) unequal economic contest between consumers and corporate power interests where electricity supply can be manipulated and price fixing is rewarded. They claim that if we just wait long enough and have enough faith, the "market" will ultimately produce a more efficient and innovative electricity system.

Therein lies the essential poverty of the very idea of deregulation: it offers no concrete vision that we can work towards as a society, only a belief that the market will decide what is best for us. It prescribes an electrical future in which we have no say, except in how we defend ourselves personally against the unregulated forces of greed. The recent Liberal and Conservative retreat from retail deregulation is only temporary. Neither of them has abandoned the view that the economic law of the jungle must prevail. What a pathetically helpless future this view offers. We go where the market takes us, adrift on a sea of forces we

"Father just turned a switch and there was light. I'll never forget it."

In 1996, for its 90th Anniversary, Ontario Hydro ran ads evoking the spirit of public power and the great impact it had on the province. Ironically, this was also the year the Conservative government began making plans to fragment and sell off Hydro and thereby destroy the public system that had contributed so much to Ontario's prosperity. New Democrats are unalterably committed to preserving the best aspects of our public power heritage and reshaping it, where necessary, for the twenty-first century. *Hydro One Archives*

cannot command.

We can do better than this. Those who came before us certainly did.

I am at Niagara, downstream from the Falls, studying from a distance the Sir Adam Beck 1 Generating Station, still as impressive today as it must have been when it was built 80 years ago. Miniature waterfalls flow from the evenly spaced apertures that punctuate its broad expanse. Behind each of these openings, I know, is a mighty turbine that has already captured the energy of the falling water and delivered it to a transmission network that would, at a marathon per day, take more than two years to traverse. More impressive than these massive works, however, is the idea that gave them birth: the notion that cooperation, not competition, was the better way to light the future.

Public power is not just a system. It is a canvas on which we can paint any system that is technologically possible and that meets our needs as a society. As we acquire new pigments—new technologies, say, or new uses for our power—we can change or add to our picture because *we* hold the brush, not some impersonal, uncaring collection of market forces. It is certainly possible that some of our brushstrokes and figures will be flawed. It is predictable that we will occasionally wake up and look with disappointment at some aspect of yesterday's efforts, some element in the painting that doesn't work as it should, and say to ourselves: "We can do better." As long as we hold the brush, though, we are not stuck permanently with our lesser judgments. We can learn from our mistakes and, even though it may take a while, eventually paint them over. Even Rembrandt improved with time and reflection.

I share with many people, and I hope you as well, a vision of Ontario's future public power system that literally empowers us all. Every home, every business, every farm, every public building will be able to generate clean, renewable energy, and share it with others, at a cost that steadily declines. Orchestrating this symphony of sources will be public agencies, municipal and provincial, that own and maintain our networks, as well as the central power stations that will continue to be needed to ensure reliability and to support economic development.

We already possess the technology to paint this picture. It will take time, but we have the time, provided we have the will. It will take money, but only the money that we are already spending, more wisely and efficiently allocated. Most importantly, our painting must have a perspective, a point of view that guides our hand as our picture evolves. And I can think of no better guide to our energy future than the stirring motto of the now-dismantled Ontario Hydro, words that were most certainly chosen by Adam Beck himself as the best expression of the vision of the public power movement he so ably led: *Dona Naturae Pro Populo Sunt*—the Gifts of Nature are for the People.

Notes

Chapter 1

Honeymoon Capital: Local legend has it that Niagara Falls became the honeymoon destination of choice when Napoleon's brother, Jerome Bonaparte, travelled there by stagecoach from New Orleans for his honeymoon and returned with glowing reports. **Lord Dufferin:** While several sources were consulted about the early history of Niagara Falls developments, the author is most indebted to Merrill Denison's *The People's Power*, chapters 1-3. **Nikola Tesla:** For an engaging story of Tesla's life and achievements, see *www.pbs.org.tesla*. **Horsepower:** Electrical energy is now universally measured in watts and its increments (kilowatt, megawatt, gigawatt and terawatt). One kilowatt is the energy needed to light ten 100-watt lightbulbs. One megawatt (1,000 kilowatts) equals 1,340 hp. Therefore 10,000 hp equals about 7.5 megawatts. Ontario's current summer peak demand of about 25,000 megawatts translates to 33.5 million horsepower. **Canadian Niagara Power Company:** The company survived and eventually built a generating station named after founder William Birch Rankine. The 74.5 megawatt station is still operating. See: *www.cnpower.com*. **William Mackenzie:** The best biography of Mackenzie is R.B. Fleming's *The Railway King of Canada,* but there are also fascinating accounts of Mackenzie in Pierre Berton's *The National Dream* and *The Last Spike*, the epic story of the building of Canada's first transcontinental railroad. **Henry Mill Pellatt/Casa Loma:** You can take a virtual tour of Casa Loma and learn more about Pellatt at *www.casaloma.org*.

Chapter 2

Ellis/McNaught: P.W. Ellis was a jeweller and W.K. McNaught was a watchmaker. McNaught later went on to represent North-west Toronto in the Ontario Legislature and was also an author and poet. **Snider/Detweiler:** Denison, pp. 32-35. **Annie Taylor:** One story is that Taylor, an impoverished schoolteacher from Bay City, Michigan, was desperately contemplating suicide by jumping into the Saginaw River. It was then that she came up with the idea of a money-making publicity stunt in which she would become famous or die trying. She lived to age 86 and died in 1923, flat broke. See *http://bay-journal.com*. **Adam Beck:** The most comprehensive biography of Adam Beck was the first one ever written on Ontario Hydro's founder: W.R. Plewman's *Adam Beck and the Ontario Hydro*. Much of what was subsequently written about the early years of Hydro was taken from Plewman's long and detailed work. **"A government Commission:"** Quoted in Denison, p. 34. **Sufficient competition to keep prices low:** In a telling historical parallel, the current Conservative government claimed for years prior to opening the electricity market in May 2002 that Ontario had plenty of surplus power, a comforting fact that would keep market prices low and service reliable. Within weeks after the Ontario market opened, the province was short of power and had to, on several occasions, import very high-priced energy from the U.S. Even domestically-generated power regularly spiked to hundreds of dollars per megawatt hour (MWh), often five to ten times higher than regulated pre-competition rates of about $40/MWh. **"Predatory and parasitical promoters:"** Quoted in Denison, p. 37. **"Guarantee electric power:"** Quoted in Plewman, 36. **"The building and operating as a public work:"** Quoted in Denison, p. 40. **Ontario Power Commission:** Freeman, *The Politics of Power*, p. 17. **"A smashing majority."** Whitney's Conservatives were the most popular governing party in Ontario history, undoubtedly because of their commitment to public power. In the 1905 election, they captured 70 percent of the seats in the Legislature. In 1908, 81 percent; in 1911, 77 percent; and in 1914, 76 percent. *For historical results of Ontario elections, see www.electionsontario.on.ca/Results.* **"The sport and prey of capitalists:"** Quoted in Denison, p. 46.

Chapter 3

Beck report: Denison, p. 49. **Government backed utilities can borrow money at more favourable rates:** Since the Enron implosion in 2001 and the subsequent exposure of widespread deceptive accounting schemes in the U.S. power trading business, lenders have been reluctant to finance projects of investor-owned utilities at anywhere near the preferential rates they used to command. This gives publicly owned utilities an even greater economic edge, as the gap between their financing rates and those of the private sector widens. Moreover, as this book goes to print, there are no known cases of a publicly owned utility engaging in questionable accounting practices in energy trading. No claims are being made here, or anywhere else in this book, that public sector financing or service operations are somehow immune to corruption of whatever sort. But the probability of a public sector entity engaging in the sorts of accounting shenanigans that Enron and other power companies have done is decidedly lower. This is true for many reasons, but if recent history is any guide, it would appear that the single most critical driver is that the executives of private concerns have what at times appears to be an almost irresistible incentive to manipulate their stock prices for their own "insider" gain. And that is what shady accounting can do very effectively, at least for a while until the bottom finally drops out. Public companies have bonds to be sure. But they don't have even the temptation of stock values to manipulate. **City Hall resolution:** Quoted in Plewman, p. 48. **"We want cheap power and we want it now!:"** Quoted in Denison, p. 51. Had the rally been held in our time they would no doubt have adopted the more modern protest cadence: "What do we want? Cheap Power! When do we want it? Now!" We may yet hear these very words at the doors of the Legislature from a reborn public power movement. **Canvassers employed by the Syndicate:** Private power companies, like U.K.-owned Direct Energy, resumed door-to-door canvassing in our time, not seeking votes, but rather long-term power supply contracts that were allowed by deregulation. Typically the contracts called for hydro rates that were more than a third higher than pre-deregulation rates. Many signed such contracts out of fear, which was stoked by the sellers, that deregulated rates would skyrocket. **Women would not have the right to vote:** There were some exceptions. "Widows and unmarried women" who met the "property or income qualification" had been granted the right to vote in Ontario municipal elections in March, 1884. (Ontario Statutes, 1884 c.32, ss.3-4). **Transmission rights-of-way:** Think of it as an unintended equity bonus for these farmers, who no doubt needed the money, paid for over time by electricity consumers. **Ontario High Court of Justice decision:** Denison, p. 78. **"Repealed the Magna Carta.:"** Quoted in Denison, p. 80. This particular gem came from an article in the *Financial Times* of London, England, which went on to say: "Such an outrageous parody of law-making must inevitably provoke energetic resistance. How a Conservative Legislature and Scotch to boot, could ever have been cajoled into passing it is a psychological puzzle. The only explanation conceivable is that Ontario is having a very severe fit of municipalizing mania. In order to get a power plant 'of the people's own' it is prepared to run any financial risk, to enter into any kind of contract, and to declare legal any kind of illegality. This all means a lovely time for the lawyers, but a bad time for investors. Manifestly the whole question must be thoroughly threshed out and satisfactorily settled before any important Ontario loan, whether Provincial or Municipal, can be offered again in London." **Ottawa would not disallow the Ontario Act:** Denison, p. 87. Freeman, *The Politics of Power*, p. 36, reports that Liberal Prime Minister Laurier himself had previously announced, on March 29, 1910, "that he would not invoke disallowance even though he found the 1909 legislation objectionable. Since the Ontario action was unjust rather than *ultra vires*, his hands were tied." **Berlin switching-on ceremony:** Plewman, pp. 65-7 provides the most evocative report of this historic event. A small correction is in order, however. The "little girl, Hulda Rumpel" in Plewman's account was actually 18-year old Hilda Rumpel, daughter of a prominent local felt manufacturer. In 1960, Hilda Rumpel Reade participated in a fiftieth anniversary re-enactment of the event staged by the City of Kitchener.

Chapter 4

"Every small village:" Quoted in Denison, p. 100. **Supply shortages:** Several times last summer (2002), Ontario was short of power and had to import as much as 4,000 megawatts, nearly 20 percent of peak demand. We paid as much as $2,000 per megawatt hour for these imports, 50 times the pre-deregulation price of power. Also last summer, a lengthy investigation by the California Public Utilities Commission found that a few large generators deliberately withheld power from the state during shortages, even during blackouts. Whatever the faults of public power might be, it is inconceivable that a publicly owned utility would harm the economy in order to keep prices sky-high. **A reasonable power surplus:** This does not mean publicly owned utilities should have *carte blanche* to build indiscriminately. Having less-than-fully-productive assets still costs power consumers, who must ultimately pay for them. But the cost/benefit calculation of how much surplus capacity they should maintain can take into account much more than the utilities' year-to-year financial statements. Environmental impacts, to name the most obvious other factor, must also be part of the balance sheet. **Power supply contracts:** Beck could get good prices out of the existing power companies because there was at this time a significant surplus of hydroelectric generation capacity in the province. As this surplus got soaked up by increasing demand, the private power owners began asking for more money, which led to the government's decision to authorize the HEPC to begin building its own hydro generating stations. This story is told in the next chapter. **Hydro Circus:** Keith Fleming refers to these travelling rural demonstrations as the "Power Circus". Plewman reports that they "were referred to humorously as 'Adam Beck's Circus.'" The *Power Workers' Union 50th Anniversary* history book called it the "Electric Circus." Denison prefers "Hydro Circus." **Lower voltages:** If electricity were water, distribution systems would be garden hoses, transmission systems would be fire hoses. The difference is actually more dramatic than this. If you hooked up your house to transmission wires and then flipped the switch, your house would instantly burst into flames. Transformers act as bridges between the two different wires systems by increasing or decreasing the voltage (pressure) of the electricity as it moves from one system to another. "**Make life on the farm more attractive:**" Quoted in Denison, p. 107. **Flat rate power:** Uniform (flat) rates for network services are often called "postage stamp" rates, after the idea that a letter mailed to your neighbour next door costs the same as one sent thousands of kilometres away. Almost all distribution systems today offer flat rates, but these rates vary by customer class. Residential, commercial and industrial are the main customer classes but there are typically many subclasses within them. These rates are designed to either promote or avoid cross subsidization between, for example, residential and industrial service costs. **Uniform Hydro rates:** These subjects are discussed in noteworthy detail in Fleming's *Power at Cost*, a must-read for those interested in Ontario's rural electrification. **111-seat Legislature:** Today's Legislature has only 103 seats, reduced by the Harris Conservatives from its previous 130. This means each Member of the Legislature now represents an average of about 107,000 of Ontario's 11 million people. In 1919, the first provincial election in which women had the right to vote, each Member represented fewer than 25,000 of the province's 2.5 million people. **Rural Electrification Administration.** A 12-minute historical film of the creation of the REA, *When the Lights Came On*, is available online at *www.usda.gov/rus/electric*.

Chapter 5

Lower interest rate: If you think my example of a 3 percent spread between government-guaranteed and private sector loan rates is exaggerating the actual difference, put yourself in the position of the banks and investors that lost tens of billions of dollars in the Enron collapse, the Pacific Gas & Electric bankruptcy, and the British Energy meltdown, to name only the best-known examples of the utter failures of deregulation and privatization. Everywhere in the world, private power companies, even the largest, are now having a difficult time getting reasonable financing rates for new merchant generation projects. Hundreds of such private projects have been cancelled or put on hold over the last two years, including some planned for Ontario. **Upstream Chippawa:** The Niagara River flows from south (Lake Erie) to north (Lake Ontario), which is why Chippawa is upstream even though it is south of Niagara Falls. **Assured of being highly profitable:** Just like today, which is why the orgy of private power construction long predicted by the Harris/Eves Conservatives as a "natural" consequence of their privatization and deregulation experiment has failed to materialize.

Chapter 6

Gregory Commission: The Commission gave private power interests another opportunity to try and discredit public power. Plewman, p. 335, recounts the "petty and personal" nature of some of the testimony given to Gregory, including the charge that Hydro had allowed the use of liquor in the Chippawa construction camps during prohibition. **Debate at City Hall:** Plewman's depiction, pp. 315-19, of the seven-day radials debate at Toronto City Council in August, 1922 is a fascinating account of clashing personalities and invective-filled debate. For more on the radials story, see also Denison, 137 *ff* , and Freeman, 50-54. **His ailing wife:** By all reports, Lady Mary Pellatt was a generous and empathic person. Among other things, she was a founder and First Commissioner of Canada's Girl Guides and used to bring large groups of the girls to Casa Loma regularly, allowing them to roam freely. **"Forge a band of iron:"** Quoted in Denison, p. 158.

Chapter 7

Adam Beck himself: Plewman, 216-20. Plewman also recounts Beck's anger at attempts to morally compromise him while he was in Washington. "He had no doubt that his enemies in the private power ring had set traps for him." **"The greatest monopolistic corporation:"** Quoted in Geist, p. 114. Nebraskan George W. Norris was one of the great champions of progressive causes in the U.S. Congress for 40 years (1903-43). He is widely regarded as the most important political force behind the creation of the Tennessee Valley Authority and was the singular proponent of the unicameral (one house) legislature in the U.S. Nebraska was the only state to adopt a unicameral form of government. **Penniless at the end:** In yet another parallel to Pellatt's life, Insull had a great mansion (now known as the Cuneo Museum) built near Chicago in 1914, the year Casa Loma was completed in Toronto. Insull lost it in his own bankruptcy a few years after Pellatt lost Casa Loma. **Securities and Exchange Commission:** On October 9, 2002, the U.S. Senate Governmental Affairs Committee declared that the SEC's oversight of Enron was "a catastrophic failure". **"A beacon light to the people of the United States:"** Quoted in Denison, p. 163. **Charles Evans Hughes:** Quoted in "New York Power Authority: Our History." At *www.nypa.gov*. **"Must rest forever in the people:"** This and following Roosevelt are taken from "A Campaign Address on Public Utilities and Development of Hydro-Electric Power," given at Portland, Ore. on September 12, 1932. See *http://newdeal.feri.org*.

Chapter 8

Beck had urged American legislators: Plewman, p. 219. **Hepburn/Beck apple incident:** McKenty, p. 14. **Henry scathingly called Hepburn:** Denison, p. 201. **Hatred of industrial unions.** In 1937, Hepburn recruited a private army, known variously as "Hepburn's Hussars" or "Sons of Mitches" to terrorize the newly organized United Auto Workers at General Motors in Oshawa, who were on strike for a first collective agreement. It didn't work. The workers, who had taken several pay cuts during the Depression, even as General Motors' profits soared, were ultimately successful. For the full story, see Pierre Berton's *The Great Depression 1929-1939*. Toronto: McCLelland and Stewart, 1990.

Chapter 9

With the support of the Liberals: On 2nd reading of Bill 35, the Energy Competition Act, 1988, S.O. 1988, c.15, every Liberal present in the Legislature voted in favour of the bill. Every New Democrat voted against it. Liberal Leader Dalton McGuinty, in particular, was a flagrant and well-documented supporter of privatization and deregulation until, in late 2002, public anger over rapidly rising hydro rates forced him to retreat temporarily and support the Conservative government's panic-driven rate freeze. **50 to 55 cents per thousand cubic feet:** MacDonald, p. 289. Gas prices can be expressed as cost per volume or energy units. In the Imperial system, this is an easy conversion. One thousand cubic feet (Mcf) of natural gas yields one million British Thermal Units (MMBtu) of energy. Convert that to metric and you have slightly more than one gigajoule (GJ). You start needing a calculator when measuring in metric volumes, such as the commonly used one thousand cubic metres ($10^3 m^3$), which gives you 35.3 MMBtu or 37.07 GJ. Got it? And that's just for the natural gas commodity. Adding in the different transmission, distribution and storage costs, which vary by province and region, further complicates the calculations. For the consumer, the most meaningful price is the burner tip cost, all the pricing elements added up and expressed as easy-to-understand dollars and cents. **Manufactured gas:** Manufactured gas was made primarily from coal. The process creates non-biodegradable tars that contaminate groundwater. There are thousands of still-toxic gas manufacturing sites in the U.S. that represent an enormous environmental challenge. It is a very good thing that natural gas has replaced manufactured gas. **A private domain:** See MacDonald, pp. 289 *ff* for the entire absorbing story of how the natural gas transmission and distribution industry was privatized by deliberate political design. **The most expensive option:**

Frost later denied that Ontario had contributed to the north of Superior line. MacDonald, p. 295. **C.D. Howe unseated by the CCF:** One of Howe's pet projects, the Avro Arrow, was spitefully cancelled by Diefenbaker (the "Chief") after the 1958 election in which he won the biggest Parliamentary majority in Canadian history. Having stood up for Canadians in the pipeline debate, the Chief destroyed their aviation industry. **Space and water heating:** Natural Resources Canada, *Improving Energy Performance in Canada*, Appendix 2. **Live Better Electrically:** The first public use of this slogan was by General Electric in its "Live Better Electrically" campaign launched in the U.S. in 1956. **Gold Medallion Home:** This idea was also an American import. But at least Ontario Hydro didn't adopt the Edison companies' anthropomorphic electricity icon "Reddy Kilowatt." **Declined for the second year in a row:** *Ontario Hydro 1995 Annual Report*, pp. 5-7. **Conservative Chairman:** The Chairman was William A. Farlinger, a high-profile Conservative fundraiser and Mike Harris golf partner who was appointed later in 1995, after the Conservative win. He would presumably be one of the last people to admit that the NDP stewardship of Ontario Hydro was paying off, but the facts were clear. **NDP energy policy:** *A New Democratic Ontario: Programme of the Ontario New Democratic Party (1971)*, p. 49.

Chapter 10

Heavy water systems: Heavy water is the common name for deuterium oxide (D_2O), which is 10 percent heavier than H_2O. Those interested in learning more about the differences between light water and heavy water reactors will find a wealth of information on the Internet. **Lengthy construction delays:** McKay, p. 66. McKay's well-documented and thoroughly damning indictment of Ontario Hydro's economically and environmentally disastrous nuclear and fossil programs is as compelling today as it was when first published in 1983. The author owes much to this highly recommended book, which he believes should be updated and republished. **Bruce A boiler contracts:** McKay, p. 69. **Dipped below $10:** Leadbeater, p. 10. This paper is available on-line at *http://inord.laurentian.ca/pdf/1a19.pdf*. **Still on the drawing boards:** This is a habit today's Conservatives also display. Former Harris-era Conservative Energy Minister Jim Wilson was especially fond of trumpeting private sector power projects that never materialized. The two Sithe 800 MW cogeneration plants near Toronto, for example, were announced more often than an Elizabeth Taylor wedding. They were finally called off in 2002. **Improved occupational health and safety law:** In 1979, Ontario workers were finally given the right to refuse to perform work that they believed to be hazardous to themselves or other workers. Prior to this, employers had the right to fire workers who wouldn't "follow orders" despite their health or safety concerns. The law also required joint labour/management health and safety committees in most workplaces. The NDP, now under the leadership of Michael Cassidy, had made passage of the law its highest priority, and used its leverage in the minority government to ensure that it was not watered down. A number of Conservatives decried the law, predicting that many businesses would desert Ontario in disgust and that those staying behind would be quickly rendered uncompetitive. **Order to start Darlington:** According to MacDonald, p. 308, Darcy McKeough told him that the final approval to start Darlington never came before Cabinet. Davis had presumably given the order himself. **One more nuclear station:** Porter said that "Ontario Hydro should base its system expansion plan on a growth range for peak capacity to the year 2000 of 2.5 to 4.0 percent per annum." At the higher end, two Darlingtons (or the equivalent in conservation or alternative generation) would most definitely have been required. This was not an entirely unreasonable assumption at the time. Peak demand growth between 1960 and 1975 had averaged 5.2 percent, although it had slowed considerably by the end of that period. Over the following 20 years, however, Porter's projections proved high. Peak demand growth averaged only 2.5 percent *(Electric Power Statistics, Volume 1, Statistics Canada, catalogue 57-204)*, which could have easily been absorbed through a steady program of energy efficiency and renewables, as the NDP had proposed. **$4 billion could have been saved:** Dissent by New Democrats, Report on Darlington Nuclear Generating Station, Ontario Legislative Assembly, Select Committee on Energy, 1985, 1st Session, 33rd Parliament, 34 Elizabeth II. **Deserved public support:** In November 2002, in front of the great symbol of public power, Niagara Falls, Conservative Energy Minister John Baird announced the Eves government's new plan to increase energy supply, since the old one, deregulation, wasn't working. It included "tax holidays" for renewable power. **Panel to review nuclear safety and emergency plans:** This was done. In 1988, Prof. F. Kenneth Hare reported that "The Ontario Hydro reactors are being operated safely and at high standards of technical performance." But he also warned: "Complacency is the Achilles' heel of large, self-contained bodies such as Ontario Hydro." In light of subsequent developments, recounted in Chapter 15, it was clear that Hydro had not heeded this warning. **Cost overruns on SkyDome:** Original estimate of SkyDome's construction costs: $150 million, of which $30 million was to come from the provincial government. Final cost: $609 million, most of which was owed by the provincial government, thanks to Peterson's generosity. In 1994, Premier Bob Rae unloaded the money-losing facility for $151 million. It had been bleeding millions of dollars a year, even during the 1992-93 seasons, when the Toronto Blue Jays won back-to-back World Series. But even wiping out its debt didn't save SkyDome, whose new owners declared bankruptcy in 1998. It sold again in 1999 for $110 million. Peterson remains a prominent figure in Toronto's professional sports circles. **Painlessly reducing energy demand:** Passmore Associates, "Electrical Efficiency Opportunities for Ontario." **Unexpectedly went down:** The shutdown happened in late spring, 2002, at one of the Bruce B units that had been leased, for a pittance, by the Harris government to British Energy-controlled Bruce Power. The shutdown was downplayed as a minor event in an innocuous press release from the company, which did not indicate that the 840 MW unit would be offline throughout the summer. The Eves government has repeatedly rebuffed NDP demands that generating unit shutdowns be publicly posted, as is required under U.S. FERC regulations. The idea is that all market players should have equal access to information on the state of the electricity system. Despite its claim that it believes in "a level playing field," the Eves government plainly doesn't.

Chapter 11

A bad time to be in government: Premier Bob Rae's memoir, *From Protest to Power*, written shortly after he left office, unflinchingly reports on the severity of the recession his government inherited from the Liberals and the often difficult choices he and his Cabinet had to make as a result. **The project would pay for itself:** Because there is a high upfront cost to retrofitting most buildings, especially older ones, it is normally done on borrowed money that is typically arranged by the retrofitting company. The usual arrangement is that building owners pay nothing upfront for the retrofit. Their only obligation is to pay the retrofit company the same amount the buildings used to incur in energy bills. The retrofit company then pays the utility and the lender out of this amount and presumably makes some profit for themselves as well. Once an agreed-upon repayment point is reached, 100 percent of the savings revert to the building owner. Alternatively, some retrofit financing agreements call for a "shared savings" contract. Under this arrangement, the building owner gets some of the benefit of reduced energy costs from the beginning. This obviously extends the repayment period and therefore the total cost of financing the project, but it does produce an immediate positive cash flow for the building owner, in addition to having a renewed building. In the last year of our government, a consortium of union pension funds led by the Canadian Commercial Workers Industry Pension Plan offered to finance tens of millions of dollars in government building retrofits. They had "patient money" to invest and were interested in job creation. After the 1995 election, the Conservative government said it wasn't interested in union pension money. Eventually, it gave the job of retrofitting a few government buildings to a company owned by SNC Lavalin, a Quebec engineering firm that also profited immensely from the Conservatives' sale of Highway 407. **Energy efficient refrigerators:** In 1989, NDP Energy Critic Brian Charlton proposed giving a new, Canadian-made high efficiency refrigerator to every household in the province, in return for their old one. It had to be a "one-for-one" program so that the old fridge didn't end up in the basement keeping a few bottles of beer and a wilting cabbage head chilled. Removing and neutralizing the ozone-destroying CFC's in the old appliances could be better controlled through a standardized program, as could the recycling of the scrap metal. Charlton's proposal would create thousands of manufacturing jobs in Ontario and would lift a large power burden from the province. It would actually be much cheaper, he calculated, to mass produce and give away efficient fridges than to build another Pickering-sized reactor. Do the same thing for water heaters and you could also close down a fair-sized coal unit. Charlton was widely ridiculed in the media for this scheme, although it made perfect economic sense, even if you did not take the thousands of extra jobs into account. Furthermore, the "free fridge" idea itself wasn't totally original; he just introduced it to Ontario, based on similar concepts from the U.S. and Europe. But the initial negative reaction was too much to overcome and the idea died. Moreover, once we became government, Darlington was already in operation and even though hydro rates were rising as a result, there was plenty of power available. That's why we decided to impose a stricter standard on the marketplace and let time take its course. **$33 trillion annually:** Wilson, p. 106. **Environmental Commissioner website:** *www.eco.on.ca*. **Maurice Strong:** Rae, pp. 230-233. **Cumulative losses of $8.3 billion:** *Ontario Hydro 1997 Annual Report*, p. 69. **Greenhouse gas emissions management strategy:** *Ontario Hydro 1994 Annual Report*, p. 17. The 1994 Annual Report was a lucid and, at times, almost inspiring vision of the path Ontario Hydro was taking to become a global utility leader in energy efficiency and sustainable development. It offers a glimpse at what might have been, until the Conservatives imposed their own view of the future of Ontario's electricity system. **Hydro debt at $31.4 billion:** I actually heard Ernie Eves, in the Legislature, attribute the 1995 decline in Hydro's debt to the election of the Conservatives in mid-year. This is quite a stretch. The fact is, Hydro's 1995 rates (frozen), operational budgets and programs were in place by the fall of 1994 and the Conservative's Farlinger didn't take over as Chairman until early November in 1995.

Chapter 12

Military coup in 1973: The U.S. did not want to lose American corporate control over the country's vast mineral resources and also needed a "stable" client state in the southern hemisphere. **Held the remaining 30 percent:** Scottish Power and Northern Ireland Electricity were also treated in essentially the same way as the generators in England and Wales. The restructuring was U.K.-wide. **North Sea:** North Sea oil and gas activity and exports were far and away the biggest contributors to Britain's economic boom during the Conservative years. The political stripe of the government had less to do with the new prosperity than the gifts of nature that came from beneath the sea. **Coal at about 35 percent:** *Conclusions of the Review of Energy Sources for Power Generation*, p. 30. On-line at *www.dti.gov.uk/energy/publications/whitepapers/review_sources*. **Don't credit privatization:** I was tempted to say that moving from coal to gas is a no-brainer. Then I realized that I sit across the floor in the Legislature from people who oppose an accelerated conversion of Ontario's coal plants to natural gas, even though we have to import both coal and gas. At least the natural gas is "imported" from some other part of Canada. Moreover, the Canadian natural gas industry buys a lot of supplies from Ontario. **40 percent less than in 1998:** The website of the U.K. regulator, Office of Gas and Electricity Markets (Ofgem), is at *www.ofgem.gov.uk*. **Cheap North Sea natural gas:** A study by the Canadian Energy Research Institute shows that lower electricity prices in the U.K. in the first decade of deregulation were in large part due to the substitution of cheap North Sea natural gas for expensive coal as an electricity fuel source: Eva Cudmore and Mitch Rothman, *The Economic Implications of International Power Sector Restructuring*, March 2000. **Winston Churchill on public ownership of railways:** Quoted in "The Great Train Robbery" by Sabby Sagall. *Socialist Review*, April 1996. On-line at *http://pubs.socialistreview index.org.uk/sr196/sagal.htm*. **Cost nearly $20 billion:** John Barber, "Klein and Eves stumble in haze of own making," *Globe and Mail*, October 25, 2002, A18.

Chapter 13

Samuel Insull: Richard F. Hirsch, "Powering a Generation," Smithsonian Institute. Hirsch uses 1992 dollars as his index for comparison. On-line at *http://americanhistory.si.edu/csr/powering/past/h1main.htm*. **Generated by coal-fired plants:** Glen Hiemstra "The Future of Energy," *Futurist News*, January 2002. On-line at *www.futurist.com/portal/future_trends/future_of_energy.htm*. **Getting government off the backs of the people:** See, for example, the websites of the Consumer Federation of America (*www.consumerfed.org*) and Public Citizen (*www.citizen.org*). **Federal deposit insurance:** One of President Franklin D. Roosevelt's early acts was the June 1933 creation of the Federal Deposit Insurance Corporation. FDIC provided a federal government guarantee of customer deposits in banks and savings and loan associations, up to a legislated maximum which was periodically raised. This was intended to restore confidence in the shattered, previously unregulated banking system. The FDIC, in effect, made taxpayers liable for bank and S&L failures. In 1980, Congress greatly increased the insurance on individual deposits from $40,000 to $100,000, even though the typical savings account at that time was only $6,000. S&L lobbying was behind the unprecedented jump. **Never served a day in jail:** Cara Degette, "Public Eye," *Colorado Springs Independent*, May 1, 2002. For his role in the $1 billion collapse of Silverado Banking in Denver, Neil Bush was fined $50,000 and barred from the banking business. He is now the founder of Ignite!, a company that produces on-line multimedia educational curriculum, which stands to benefit from the current Bush administration's efforts to pump federal money into private education companies. It pays to have the right genes. **Over $200 billion:** Charlie Higley, "Disastrous Regulation," *Public Citizen*, December 2000. On-line at *www.citizen.org/publications*.

Chapter 14

Fully embrace competition: As it turned out, Rhode Island, the smallest state, was the first to open up retail competition on January 1, 1998. California's opening was delayed three months because of software problems. **Innovative products and services:** We've heard this same nonsense here in Ontario about how competition was going to deliver "a whole new world of energy services." What are they? Where are they? And exactly why couldn't they be delivered by a publicly owned and regulated system? **$7,500/MWh:** In Canadian currency, this is more than $10,000/MWh. At those prices, it would cost over $1.00 an hour to keep a 100 watt lightbulb lit. Deregulation advocates would call this "giving the consumer clear price signals." (See Chapter 17.) **$9,999/MWh:** Some observers later theorized that power marketers were experimenting with techniques for driving up prices by gaming the market. An "expert on energy pricing" makes this precise point in

"How California got burned: The state electricity system is in shambles, and the worst may be ahead. How did things get to this point?" in the Sacramento Bee, May 6, 2001. **Customers not seeing the benefits of deregulation:** From *Utility Restructuring Weekly Update*, July 31, 1998. Published by the U.S. Department of Energy, Energy Information Administration, Washington, D.C. On-line at *www.eren.doe.gov/electricity_restructuring/weekly.html*. **1965 U.S. Northeast blackout:** Memorialized in the 1966 Doris Day comedy "Where Were You When the Lights Went Out?" The three-day blackout was traced back to a single faulty relay switch at the Sir Adam Beck 2 station. **"Don't have the same incentives to cooperate.:"** Quoted in "House bill brings reliability mandate." *Electric Light and Power*, November 1999. On-line at *http://uaelp.pennnet.com*. **In the wake of deregulation:** Wallace Roberts, "The Dimming Down of America." *The American Prospect. On-line* at *www.prospect.org*. **Out of state producers:** BC Hydro nearly drained its vast water reservoirs selling power to California during this period, though it may never collect on all its bills. When the dust settled, BC Hydro was accused of gouging Californians, tarnishing the generally positive image of Canadians in that state. **Power shortages had been non-existent:** Brennan, p. 20. **Calculated at $45 billion:** Weare, p. 3. **A colossal failure:** This was the 2001 finding of New York's RKS Research & Consulting, which conducts marketing research and customer opinion polling for the electric industry. See "Turn back the clock? Major business customers find few benefits to date from electricity deregulation," February 28, 2002. On-line at *www.rksresearch.com/press022802.html*. RKS's results were verified by the 2001 Deloitte & Touche Consumer Awareness Survey of Electric Deregulation in the USA which found that most consumers expect electricity rates to increase rather than decrease under deregulation. "Consumers appear to be more pessimistic about the benefits of electric deregulation," said the survey report's author. There are no signs that Americans have changed their minds since then. On-line at *www.deloitte.com*. **State-decreed rate cuts:** Cooper, p. 6. **Australia and New Zealand.** In New Zealand, which once had a non-profit, publicly owned electricity system resembling Ontario's, privatization, deregulation and "corporatization" have resulted in higher prices and blackouts. From February to May, 1998, the entire central business district of Auckland, the largest city, was blacked out because the private power company had failed to attend to proper maintenance of the main transmission feed. Power price increases in recent years have accompanied the sell-off of public electricity assets to companies like Canada's TransAlta, also a player in Ontario's power industry. For a detailed study of the New Zealand disaster, see "The Privatisation of New Zealand's Electricity Services" by Bill Rosenberg and Jane Kelsey (September 1999) at *www.canterbury. cyberplace.org.nz/community/CAFCA/publications/Electricity.* Deregulation is also credited with a significant increase in greenhouse gas emissions from Australia's power sector. See: *www.energymarketreview.org/submissions/ACFsub.pdf.* Deregulation in Australia has proceeded on a state-by-state basis, with very mixed results. Prices were falling before deregulation, but have become highly volatile since. When full retail competition was introduced in the state of Victoria in January 2002, suppliers increased their rates by up to 26 percent. See *www.ertf.energy.wa.org.gov.au/second_round_submissions/UnionsWA.pdf.*

Chapter 15

Reduce our $35 billion debt...: The 1994 Ontario Hydro Annual Report (p.54), which had been released before Strong's speech, puts Hydro's long term debt at just under $33 billion. The 1995 Report (p.54) shows the debt at $31.4 billion, down more than $1.5 billion in one year. **Not exactly the basket case Harris had been portraying:** Strong also made sweeping proposals for restructuring Ontario's electricity system in this speech. He talked about deregulation being the wave of the future and even toyed with the idea of privatization. At the time of his speech, it seemed like deregulation was certain to triumph south of the border and actually bring down electricity rates. Strong explained that even though Hydro's average rates were quite low, they might eventually be undercut by American generators dumping cheap coal power into the province. **"Neighbouring U.S. states."** Farlinger, p. 4. **"Bar us from open access.:"** Farlinger, p. 4. Ontario has long exported power to the U.S., but could only sell to neighbouring utilities, not directly to customers. Open access means that Ontario Power Generation can market directly to U.S. customers, though it hasn't been doing much of that recently. The flow of electricity since deregulation has, for the most part, been south to north. **Enlarging the NDP's rate freeze:** The NDP rate freeze was for three years, from 1993-95. After that, rates were not to be raised any higher than the inflation rate. The Conservatives then froze rates until 2000, which meant that real rates declined because they were not keeping pace with inflation. Since Hydro's input costs were not frozen, however, the money had to come from somewhere. It was added to the Hydro debt. **"If it's a problem, I'll resign:"** Toronto Star, February 12, 1996, p. A-14. **Acres International:** Acres, an internationally respected Canadian company, was founded in 1924 by Henry Girdlestone (Harry) Acres, one of Adam Beck's "young whippersnappers" who had correctly estimated the cost of building a public transmission line from Niagara to Toronto (see Chapter 3). Acres, a brilliant power system engineer, was a key figure in the Queenston-Chippawa project and went on to build an even larger hydroelectric station in Quebec, 20 years later. It made sense to have someone from Acres either on the Macdonald Committee or advising it. **Niagara stations kept in public hands:** At the June 1996 press conference for the public release of the report, Macdonald was asked why he had advised keeping the Niagara system assets under public ownership. "It was purely political," he responded with a wry smile that signalled his less than enthusiastic support for the recommendation. It was a tacit admission that the PWU campaign had had some effect on the government. **Bay Street law firms:** For a comprehensive look at how Toronto's major law firms were hoping to cash in on deregulation and privatization activity, see Bill Reno's three-part series in *LEXPERT: The Business Magazine for Lawyers* in May, June and July, 2000. **Atomic Energy Control Board:** The name of the AECB was changed in May 2000 to the Canadian Nuclear Safety Commission (CNSC). **A leading U.S. nuclear expert:** Martin Mittelstaedt, "Ontario Hydro may shut nuclear plants," *The Globe and Mail*, August 21, 1996. **Accept low standards for equipment performance and condition:** "Nuclear Report card: Ontario's Reactors are Minimally Acceptable. A Report to Ontario Hydro Management. The IIPA/SSFI Evaluation Findings and Recommendations." July 21, 1997. On-line at *www.ccnr.org/ hydro_report.html*. **Lose out in the race for investment and jobs:** "Direction for Change" p. 15. **Agricultural workers' right to join unions:** As one of its first acts in power, the Harris government, in 1995, repealed the NDP's *Agricultural Labour Relations Act, 1994*, claiming that Ontario's farmers should not be held hostage to the threat of strikes. The *ALRA* had been aimed at large agribusiness operations, not small farmers, but in any case, the law did not allow strikes. Contract disputes were to be settled through third-party mediation and arbitration, as was the case with Ontario's highly unionized health care workers. I relate this story here because this kind of distortion and outright fabrication of "facts" is typical of the Conservatives' ruling style. We can all have different interpretations of the same facts, but the Conservatives seem to have no problem simply making things up to suit their rhetorical purpose and nowhere has this been more the case than with hydro deregulation and privatization. *Follow up:* In December 2001, thanks to the persistence of the United Food and Commercial Workers (UFCW Canada) the Supreme Court of Canada struck down as contrary to the Charter of Rights and Freedoms the Conservatives' denial of union rights to agricultural workers. The Eves government, however, passed another law, supported by the Liberals, again preventing agricultural workers from joining unions or bargaining collectively. The new law is headed back to the Supreme Court of Canada.

Chapter 16

"Generating surplus will shrink considerably:" From the text of a speech by Pat McNeill to the Toronto Electric Club, September 10, 1997. **"But it's manageable.:"** Quoted in Andrew Duffy, "Hydro may have to shop for power: 'Sizeable' shortfall seen by 2000," *Hamilton Spectator*, August 30, 1997. **Hydro Hotwire:** I began publishing *Hydro Hotwire* in June 2001. You can find all back issues at *www.publicpower.ca*. **Well over 25,000 MW:** It's actually an ugly sight when every station OPG operates is running "flat out" because that includes all of the approximately 10,000 MW of fossil generation in the province.

Alberta deregulation: Reno, "Power Lawyers." The article is mainly about the spectacular increase in energy-related legal business as a result of deregulation in Alberta. **Large industrial and commercial users:** James Stevenson, "Alberta power deregulation costs and benefits still unclear 22 months later." *Canadian Press*, November 10, 2002. **"Gauge supply and demand:"**. Ibid. **Natural gas plant in Ontario:** Just in case you think that Atco's proposed project is a sign that the private sector is showing great confidence in Ontario's power market, know that the $403 million/580MW Brighton Beach station is a joint venture with still publicly owned Ontario Power Generation. The former J. Clark Keith Generating Station on the Windsor shoreline will be converted to natural gas. A great idea, but it is doubtful it would have happened without the participation of OPG. The energy produced by Brighton Beach, assuming it gets built, will be marketed by Houston-based Coral Energy, a division of Shell Oil. The power will, of course, go to the highest bidder, on either side of the border. **The signal…is higher prices:** Stevenson, *op. cit.* **As low as 2.6¢/kWh:** On its homepage, *www.theimo.com*, displays the "Average Price Since May 1." This is a misleading view of Ontario's deregulated electricity costs, however, because it is a simple arithmetical average of hourly spot market prices. I am not suggesting that the IMO is deliberately misleading us; they are probably conforming to some common reporting protocol. But the obvious fact is that much more power is purchased during the day, when spot market prices are naturally higher, than during the night, when most of us are asleep and prices are low. The accurate figure for how much Ontario is paying for its power is found on the website by clicking on Market Summaries, then clicking on the desired week, and then scrolling down to the chart that shows Hourly Ontario Energy

Price. Now look under the table under that chart. There you will find the "weighted average" price based on Ontario demand for that week, as well as the weighted average since market opening. The truth is out there, you just have to dig for it. **"A delay rather than a permanent cancellation:"** Quoted in April Lindgren, "Hydro One is no longer for sale," *Ottawa* Citizen, January 21, 2003. p. A1. **Arthur Dickinson/Judith Andrew:** Quoted in Steve Erwin, "Caps will zap hydro market: Industry." *The London Free Press*. November 12, 2002, p. C1. **"Immediate and decisive action:"** From a November 12, 2002 Ministry of Energy news release. A news release issued the next day on alternative and renewable generation also promised "immediate and decisive action." **Hold shares in them:** It was not entirely a private sector-driven plan. Baird's announcement included a commitment to finally allow the Beck Tunnel Project at Niagara to proceed. As well, another study on the feasibility of the long-proposed Beck 3 generating station, also at Niagara, was ordered. The NDP had proposed this years ago as part of the alternative to Darlington. **So much better.** Remember Donald Macdonald's comment in June 1996 when asked why he didn't recommend the privatization of the Niagara stations: "Purely political"? That could easily change. It would probably happen like this: a Conservative government would enter into a long-term lease of the Beck Stations with a private company, just as they did with the Bruce nuclear stations. The Liberals, who believe in private power, could do this as well. Then they would claim, as Eves does now with Bruce, that they haven't really privatized the stations at all. We still own them, even though we get none of the benefits of ownership.

Chapter 17

Electricity cannot be stored: There are small ways we can work around this physical law, such as rechargeable batteries, which store potential chemical energy. We can convert electricity to heat energy that is stored in some medium, such as bricks, and discharged slowly. We can also store water in reservoirs to generate electricity sporadically, as needed. But once electricity is generated, it must be used or grounded. The power is not waiting there in the wires for us to turn on a switch. The problem of power storage is the subject of research efforts worldwide. Small, incremental gains are being made, but there are not yet any significant advances. Using electricity to separate hydrogen from water and then using that hydrogen to run fuel cells shows the most long-term promise for breaking through the storage barrier. (See Chapter 19.) **Reliability insurance:** When the NDP left office in mid-1995, Ontario Hydro had a generating surplus of about 20 percent over that year's peak, a more than sufficient surplus. **Prepared to dicker:** In ancient Rome, negotiating the cost of putting out a fire in progress was a common practice until one disastrous fire destroyed a quarter of the city. The first publicly supported fire brigades were then established by the Emperor Augustus, around the time of the birth of Christ. Predictably, the Romans soon started to complain about the publicly paid firefighters sitting around most of the time. Closer to our time, privately run firefighting enterprises were run like insurance schemes in New York and elsewhere, only putting out fires of subscribers. **Until after you use it:** Wholesale and large industrial customers have access to real-time price information through "interval" or "time of use" meters. Residential versions are available but are still very uncommon. Even if you had such a meter, however, it would be a stressful way to live, constantly running to check the price of power and making hundreds of little decisions a day on whether or not you can afford to turn up the heat, turn on the lights, watch television or take a shower. This is not progress. **Prepaid electricity metering:** A sales brochure for an Ontario manufacturer of prepaid metering systems declares: "Convenient prepaid metering systems improve profit margins and customer satisfaction." on-line at *www.oxfordmediagroup.com/downloads/omg-infoenergy-lrg-brochure.pdf.* I understand the part about improved profit margins for utilities, but I doubt that continuous anxiety about whether or not your electricity will run out before your next paycheque will lead to more customer satisfaction. **"Have a huge impact:".** Quoted in Richard Brennan, "Tories to cut sales taxes on energy-efficient appliances." *Toronto Star,* November 26, 2002. **928 pages long:** The Ontario Market Rules manual can be found at *www.theimo.com/imoweb/manuals/marketdocs.asp.* **Costs of enhancing the North American grid vary considerably:** This is one of the very big and very nasty secrets of electricity privatization and deregulation. To have "real competition", electricity generators need to be able to transmit their power further and to more places. That means building billions of dollars worth of new transmission lines. Who pays for this? You do, through big increases on your hydroelectricity bill. See: "All Pain, No Gain: Restructuring and Deregulating in the interstate electricity market," *Consumer Federation of America,* Fall 2002.

Chapter 18

"**I didn't realize it then:**" Quoted in "The New American Majority. Sarah Ruth van Gelder interviews Rep. Dennis Kucinich," *YES! A Journal of Positive Futures,* Summer 2002. pp. 44-47. On-line at *www.yesmagazine.org/22art/Kucinich.pdf.* **Eleanor Clitheroe:** Clitheroe was being paid $2.18 million annually, plus a $174,000 per year car allowance, plus a $6 million severance package and a $750,000 per year pension upon retirement. **A strategic partnership:** Quoted in "Ontario government to seek Hydro One Private Partner," *Electricity Forum,* July 2002. On-line at www.electricityforum.com/news /jul02/hydro_partner.htm. **Mississagi River:** Here's how a Brascan Fact Sheet boasts of these new assets: "The Mississagi plants are in excellent physical condition, have low operating costs, long life expectancy interconnections through high voltage transmission lines with the Hydro One network." Operating costs are pegged at 0.6¢/kWh and the average price of power sold is 5.7¢/kWh. The Fact Sheet also makes it clear that Brascan intends to export at least some of this power to the U.S. In fact, Brascan's own corporate documents show that it made a profit of $8.8 million selling the hydropower from the dams during the summer of 2002. Brascan companies had contributed over $140,000 to Ernie Eves's campaign to become leader of the Ontario Conservatives and Premier of Ontario. Not a bad return—$140,000 to the Eves leadership campaign in March, 2002, and a $8.8 million profit six months later. **Junk bond issuers:** The U.S. electric power industry is experiencing its worst credit crunch since the Great Depression and it is only likely to get worse. A recent report by credit rating agency Standard & Poor's says that in the first nine months of 2002 there were 135 credit downgrades of utility holding companies and their subsidiaries, nearly quadruple the number in the year-earlier period. Fully one-third of the major companies in the power sector are on watch for future downgrades. Rebecca Smith, "Electric Industry Hits Credit Crisis" *Wall Street Journal,* October 15, 2002.

Chapter 19

Came close to blackouts: During debate in the Legislature on December 5, 2002, I questioned Minister of Energy John Baird about keeping secret the names of power companies that suspend electricity production and increasing fears of blackouts or brownouts. "Minister, don't you think the people of Ontario deserve to know whether or not the lights and the heat will come on when they flick on the switch?" Baird responded: "People of Ontario just have to flick the switch and they'll know whether the lights come on." I later said that Baird's "cavalier" comment "indicates a Minister of Energy who frankly is not doing his job of protecting the people of Ontario and their need for heat, lights and electricity when we have temperatures of 20 below." Richard Brennan, "Energy minister under fire," *Toronto Star,* December 6, 2002. **Adam Beck's young engineers:** See Chapter 3. **Seventy percent**

oppose it: A poll released June 25, 2002 by the Ontario Electricity Coalition (OEC) showed that 71.2 percent of Ontarians were opposed to breaking up Ontario Hydro and selling to private investors; only 15.6 percent thought it was a good idea. *http://electricitycoalition.org/articles/release_062502.html*. Even in Ernie Eves's own riding, Dufferin-Peel-Wellington-Grey, a poll taken by the Ontario NDP during the Premier's by-election campaign in April 2002, found 69.6 percent opposed to the Conservatives' deregulation and privatization plans; 75 percent believed Hydro should remain in public hands. **Systems benefits charges:** "The Role of the System Benefit Charges in Supporting Public Benefit Programs in Electric Utility Restructuring." On-line at *www.energyprograms.org*. **Hydro Quebec's debt:** Rapport de Dominion Bond Rating Service Limited, 2002-07-03. HQD-7, Document 3.4. On-line at *www.regie-energie.qc.ca/audiences/3492-02/Requete3492/HDQ-07-03-4.pdf*. **"No more oil:"** Quoted in Alanna Mitchell, "A world without oil," *The Globe and Mail*, December 7, 2002, p. F7.

Sources

Books

Biggar, Glenys, Ed. *Ontario Hydro's History and Description of hydroelectric Generating Stations*. Toronto: Ontario Hydro, 1991.

Daniels, Ronald J., Ed. *Ontario Hydro at the Millennium*. Montreal & Kingston: McGill-Queen's University Press, 1996.

Denison, Merill. *The People's Power: The History of Ontario Hydro*. Toronto: McClelland & Stewart, 1960.

Ernst, John. *Whose Utility? The Social Impact of Public Utility Privatization and Regulation in Britain*. Buckingham, U.K.: Open University Press, 1994.

Fleming, Keith Robson. *Power at Cost: Ontario Hydro and Rural Electrification, 1911-1958*. Montreal & Kingston: McGill-Queen's University Press, 1992.

Fleming, R.B. *The Railway King of Canada: Sir William Mackenzie, 1849-1923*. Vancouver: UBC Press, 1991.

Freeman, Neil B. *The Politics of Power*. Toronto: University of Toronto Press, 1996

Geist, Charles R. *Monopolies in America: Empire Builders and Their Enemies from Jay Gould to Bill Gates*. New York: Oxford University Press, 2000.

Laird, Gordon. *Power: Journeys Across an Energy Nation*. Toronto: Penguin Books Canada, 2002.

MacDonald, Donald C. *The Happy Warrior: Political Memoirs*. Markham: Fitzhenry & Whiteside, 1988.

McKay, Paul. *Electric Empire: The Inside Story of Ontario Hydro*. Toronto: Between the Lines, 1983.

Plewman, W.R. *Adam Beck and The Ontario Hydro*. Toronto: Ryerson Press, 1947.

Rae, Bob. *From Protest to Power*. Toronto: Viking (Penguin), 1996.

Reno, Bill. *The Power Workers' Union 50th Anniversary: A history of the union of workers behind the construction, operation and maintenance of one of the world's great power systems*. Toronto: Power Workers' Union, 1996.

Rifkin, Jeremy. *The Hydrogen Economy*. New York: Tarcher/Putnam, 2002.

Wilson, Edward O. *The Future of Life*. New York: Knopf, 2002.

Magazines & Reports

Brennan, Timothy J. "The California Electricity Experience, 2000-01: Education or Diversion?" Washington, D.C., Resources for the Future. October 2001, 57 pp.

Cooper, Mark. "All Pain, No Gain: Restructuring and Deregulation in the Interstate Electricity Market." 38 pp. Washington, D.C.: September 2002.

Cudmore, Eva and Rothman, Mitch. "The Economic Implications of International Power Sector Restructuring." March 2002. 297 pp. Canadian Energy Research Institute.

Direction for Change: Charting a Course for Competitive Electricity and Jobs in Ontario. Toronto: November 1997. 41 pp. (Government of Ontario White Paper, Ministry of Environment and Energy.)

Electrical Efficiency Opportunities for Ontario. The Real Potential for Limiting Future Supply Commitments. A Report to Bob Rae, MPP and Brian Charlton, MPP by Passmore Associates International." July 1989. 33 pp.

Farlinger, W.A., Homer, G.J., Caine, B.S. "Ontario Hydro and the Electric Power Industry: Vision for a competitive industry; helping Ontario to thrive to and beyond 2000." 41 pp. Toronto: August, 1995.

A Framework for Competition. The Report of the Advisory Committee on Competition in Ontario's Electricity System to the Ontario Minister of Environment and Energy. Toronto: Government of Ontario, May 1996.

Hirsch, Richard F. "Powering a Generation," Washington, D.C.: Smithsonian Institution, 2001.

Improving Energy Performance in Canada: Report to Parliament under the Energy Efficiency Act, 2000-2001. Ottawa: Natural Resources Canada, 2001.

Leadbeater, David. *"The Development of Elliot Lake, 'Uranium Capital of the World:' A Background to the Layoffs of 1990-1996."* 51 pp. Sudbury: Laurentian University, July 1998.

Ontario Hydro Annual Reports: 1994, 1995, 1996, 1997.

Report of the Royal Commission on Electric Power Planning: Volume 1: Concepts, Conclusions and Recommendations. Toronto: February 28, 1980 (Porter Report).

Reno, Bill. "Restructuring Ontario's Electricity Industry: A Legal Cornucopia Awaits," *Lexpert,* Volume 1, Issue 7. Toronto, May 2000.

Reno, Bill. "The New Deregulated Regulatory Labyrinth," *Lexpert,* Volume 1, Issue 8. Toronto, June 2000.'

Reno, Bill. "Power Projects and Power Purchase Contracts, *Lexpert,* Volume 1, Issue 9. Toronto, July/August 2000.

Reno, Bill. "Power Lawyers," *Canadian Lawyer*, September 2001, Volume 25, Issue 9.

"The Role of the System Benefit Charges in Supporting Public Benefit Programs in Electric Utility Restructuring." Issue Brief for the National Association of State Energy Officials. September 9, 1999.

Sagall, Sabby, "The Great Train Robbery," *Socialist Review*, April 1996.

U.K. Department of Trade and Industry, *Conclusions of the Review of Energy Sources for Power Generation*, October, 1998.

Weare, Christopher. "The California Electricity Crisis: Causes and Policy Options." 140 pp. San Francisco: The Public Policy Institute of California, January 2003.

Index

Abitibi Canyon, 104
Acres International, 191, 277
Advisory Committee on Competition in Ontario's Electricity System, 190
AECL, 123-124
AES Corp., 239
Aird, John J., 103
Alberta, 16, 57, 110, 112, 190, 202-205, 220, 222, 224, 233-234, 239, 279
Alberta Energy Company Ltd., 190
Alberta Energy Utilities Board, 202
Alberta Power Pool, 202
Alternative sources of fuel, 165
American Electric Power, 157
American Federation of Labor, 93
Andognini, G. Carl, 194-195
Anti-Hydro Citizens Committee of Business Men, 45
Arizona, 115, 173, 211
Arkansas, 173, 176, 201
Arran Township, ON, 71
Artificial electricity shortages, 223
Assembly Bill 1890, 172
Association of Major Power Consumers, 211
AT&T, 240
Atomic Energy Control Board, (AECB), 193, 278
Atomic Energy of Canada Ltd., 123
Auckland, New Zealand, 276
Austin, TX, 232
Australia, 185, 276-277
Automakers, 166
Automobiles, 52, 85, 161
Avro Arrow, 270

Babcock and Wilcox, 124, 127
Baird, John, 45, 209, 220, 243, 271, 282
Bank of Commerce, 48

Barber, John, 274
Barrie, ON, 56, 70
BC Hydro, 189, 202, 233, 276
Beauharnois Light, Heat and Power Company, 101, 103
Beaver River, 70
Beaverton-Canning, ON, 70
Beck, Adam, 32-36, 39-45, 47-48, 50-53, 55-57, 59-61, 63-66, 69-77, 79-87, 91, 94-95, 98-100, 102, 104-105, 107, 109, 111, 118, 130, 178-179, 187, 191, 197, 212, 224, 226, 235-236, 240, 246, 261-264, 266, 268-269, 276-277, 280, 282
Beck Commission, 40-41
Beck Stations, 77, 178, 280
Beck Tunnel Project, 280
Beck's Hydro-Electric Railway Act, 83
Beck's Power Bill, 43
Belle River, ON, 63
Berlin, ON, 30, 32, 34-35, 44, 46, 50-53, 122, 265
Big Chute, 70
Bill 35, 197, 200, 269
Bill 58, 236
Biomass, 130, 137, 143, 165, 250, 254
Biomass energy, 250
Birmingham, UK, 163
Blair, Tony, 158
Bohr, Neils, 122
Bonaparte, Jerome, 263
Bonneville Power Administration, (BPA), 91, 93, 180, 189, 232-233
Boundary Water Treaty, 1909, 98
Boyd, Sir John Alexander, 49
Bradford, ON, 85
Brantford, ON, 32, 44, 46, 62
Brascan, 237-238, 282
Brennan, Richard, 281-282
Bridgeport, ON, 32
British Airways, 154

— 287 —

British Columbia (B.C.), 233
British Energy, 86, 155-156, 159-161, 201, 237, 246, 267
British Gas, 154
British North America Act, 50
British Rail, 160-161
British Steel, 154
British Telecom, 154
Brockville, ON, 55
Bronconnier, Dave, 235
Bruce nuclear station, 136, 156, 190, 199, 201, 227, 237-238, 272, 280
   Bruce A, 123-124, 127, 136, 194, 270
   Bruce B, 127, 272
   Bruce Heavy Water Plant, 127
   Bruce Nuclear, 190, 201, 237, 280
Brush Electric Light Company, 22
Buckingham Palace, 144
Budd, Peter, 211
Buffalo, NY, 24-25
Bush, George W., 91, 167, 233
Bush, Neil, 275

Cabinet, 39, 43, 50, 64, 73, 79, 97, 112, 135, 139-140, 147, 190, 271-272
Calgary, AB, (City Council), 234-235
California, 16, 57, 150, 166, 170, 172-175, 177, 179-182, 200-202, 205-206, 222-223, 232-233, 243, 247, 249, 255, 266, 275-276
California independent system operator, 179
California Public Utilities Commission, (CPUC), 180-181, 266
Cambridge University, 121
Cameron Falls, ON, 71-72
Campbell, Gordon, 233
Canada Trust, 146
Canadian Broadcasting Corporation (CBC), 139

Canadian Energy Research Institute, 274
Canadian Federation of Inde-pendent Business, 211
Canadian Manufacturers Association, 31, 34
Canadian National Exhibition, 105, 117-118, 225, 253
Canadian Niagara Power Company, 25, 43, 263
Canadian Nuclear Safety Commission (CNSC), 238, 278
Canadian Union of Public Employees (CUPE), 14, 197, 236
CANDU, 119, 123
Caribou Falls, 109
Carr, Jan, 191
Carter, Jenny, 144
Carter, Jimmy, 123, 165
Casa Loma, 26, 47, 85-86, 263, 268
Cassidy, Michael, 130, 271
Cataract Power Company, 46
CCGT, 157
Central Electricity Generating Board, 155
Centrica, 158, 160
Chalk River, 122-123
Charlton, Brian, 6, 132, 134, 136, 273
Chatham, ON, 62
Chenaux, 109
Chernobyl, 134
Chicago, IL, 90, 103, 174, 176, 268
Chicago Edison Company, 90
Chicago Mercantile Exchange, 176
Chile, 153-155
Chippawa, 73, 267-268
Churchill, Winston, 161, 274
Churley, Marilyn, 201
Cleveland, OH, 87, 229-231, 239
Cleveland Electric Illuminating Company (CEI), 229-230
Cleveland Municipal Light System, 229

Cleveland Public Power, 231
Clitheroe, Eleanor, 236, 281
Coal, 21, 32, 77, 86, 97-98, 106, 109, 111, 119-120, 122, 142-145, 164-165, 169, 173, 177, 195, 221, 240, 245, 269, 273-274, 277
   Coal industries, 164
Coal plants, 24, 70, 100, 116, 119, 124, 126, 136, 142, 145, 150,157, 166, 184, 216, 224, 226, 249, 274, 275
   Coal pollution, 195
Coldwell, M.J., 114
Collingwood, ON, 56, 70
Committee of Water Powers of the House of Representatives, 87
Commonwealth Edison of Illinois, 89
Communications, Energy and Paperworkers, 235
Competitive Transition Charge, 170, 182
Connecticut, 170
Consequences, 119, 140-142, 151, 158, 191, 218, 246
Conservative party, 103, 111-112, 154, 238
Consumer Federation of America, 184, 275, 281
Cooke, J. R., 102
Co-operative Commonwealth Federation (CCF), 106, 111-112, 114-115, 270
   Ontario CCF, 111
Cornwall, ON, 95, 98
Council on Renewable Energy, 146
Crown Forest Sustainability Act (CFSA), 147-148
Cudmore, Eva, 274

Darlington Nuclear Generating Station, 123, 127, 129, 133, 217, 271
Davis, Bill, 127, 130, 135, 190

Davis, Gray, 180, 205-206
Decatur, Clarence, 112
DeCew Falls, ON, 46
Decatur, Clarence, 112
Decommissioned nuclear stations, 110
Debt Recovery Charge, 170
Degette, Cara, 275
Delaware, 176, 201
Democrats, 19, 91, 106, 135, 147, 181, 187, 205-206, 234, 241, 261, 271
Denison Mines, ON, 124-125
Denmark, 250
Department of Water Resources, 180-181
Depression, 91, 98, 101, 105, 140-141, 269, 282
deregulation, 15-19, 41, 86, 110, 115, 117, 120, 122, 130, 135-136, 151, 153, 159-160, 163, 165-171, 173-182, 184-185, 188, 190-191, 193, 195-197, 199-211, 213, 215-216, 218, 222-224, 227, 229, 233, 239-240, 243, 247, 252, 254, 260, 265, 267, 269, 271, 274-279, 281, 283
Des Joachims, 109
Detroit, MI, 55, 62, 99
Detroit River, 62
Detweiler, Daniel B., 30
Dickinson, Arthur, 211, 280
Direct Energy, 158, 265
Dominion Government, 22, 50
Douglas Point, 123, 127
Drew, George, 104, 111
Drury, E.C., 64, 76
Dubreuil Forest Products, 206
Duchess Dowager of Kent, 109
Duffy, Andrew, 279
Duke, James, 89
Dundas, ON, 53

Ecker, Janet, 236
EDF, 157

Edison General Electric, 88, 90
Edison Mission Energy, 157
Edison, Thomas, 22, 88, 90
Edmonton Electric Light and
    Power Company, (EPCOR), 234
Efficiency Ontario, 248-249
Egypt, 164
Electrical Development Company,
    23, 34, 36, 41
Electrical Power Development
    Company, 104
Electrical Union, 93
Electricity Act, 154
Electricity Consumers' Power to
    Choose Act, 172
Electricity Distributors Association,
    63
Electricity Forum, 282
Elgin West, ON, 102
Electricity Pool, 155, 158
Elliot Lake, ON, 124, 285
Ellis, P.W., 29, 263
Energy Competition Act, 195-196,
    198-199, 236, 269
Energy Intensive Users Group, 158
ENMAX, 234-235
Enron, 16, 41, 86, 91, 157, 173-174,
    181, 184, 221, 238, 246, 264, 267-
    268
Environmental Bill of Rights, 147
Environmental Commissioner, 147,
    273
EPCOR, 234, 239
Ernst & Young Canada, 188
Essex, ON, 63, 80
Eugenia Falls, ON, 70
Europe, 61, 71, 73, 83, 154, 273
Eves, Ernie, 136, 141, 149, 184, 189,
    208, 210, 222, 236, 238, 247, 251,
    254, 273, 282-283
Exelon, 89
Exxon, 145

Farlinger, Bill, 149, 151, 200
Farlinger, William A., 188, 270
Fastow, Andrew, 86
FDIC, 275
Federal Energy Regulatory
    Commission (FERC), 91, 171,
    172, 179-180, 183-184, 189, 224,
    233, 272
Ferguson, G. Howard, 65
First Nations, 72, 107, 119, 146, 148,
    151, 191, 238
Fleming, Keith, 266
Florida, 232
Forest Renewal Trust Fund, 147-148
Fort Frances, ON, 13-14, 187, 225
Fort William, ON, 71
Fossil fuel emissions, 17
Fox, John, 150
Franklin D. Roosevelt Power
    Project, 95
Friedman, Milton, 153
Frost, Robert, 111-113, 127, 270

Galt, ON, 32, 44, 46, 53
Gates, Bill, 240
Gatineau Power Company, 101
Gatineau River, 101
General Electric, 88, 90, 232, 270
General Electric Canada, 191
General Motors, 269
Georgetown, ON, 85
Georgia, 166
Georgian Bay, 56
Germany, 33, 122, 250
Gigantes, Evelyn, 128-129
Gillespie, Robert, 191
Girdlestone, Henry, 277
GO Transit System, 85
Gould, Jay, 284
Gramme, Zénobe, 22
Grand Coulee Dam, 92
Grand Valley, ON, 30
Grant, John, 22, 191
Green Communities, 146, 249

Gregory Commission, 80-81, 84, 268
Gregory Report, 81-82
Gregory, William D., 79
Grey-Bruce, ON, 63
Grier, Ruth, 132, 134, 147
Grossman, Larry, 135
Guelph, ON, 32, 44, 46, 53, 80, 146
Guindon, Roger, 206
Guinn, Kerry, 177
Gzowski, Sir Casimir, 22

Hamilton, ON, 23, 43-44, 46, 53, 55-56, 85, 200, 279
Harris, Mike, 139, 152, 187, 251, 270
Harris, Walter, 115
Hearst, William H., 71
Hearst, Sir William, 64, 70
Hendrie, John S., 43
Henry, George S., 101
Hepburn, Mitchell F., 101, 193, 251
Hespeler, ON, 32, 44, 46, 53
High Court of Justice for Ontario, 25
Higley, Charlie, 275
Hirsch, Richard F., 274
Hogg, Thomas, 104
Holden, Otto, 109
Hoover Dam, 92
Hoover, Herbert, 91
Howe, C.D., 112-113, 115, 270
Howland, O. A., 31
Hughes, Charles Evans, 93, 268
Hydro Board, 194
Hydro Circus, 55, 59-60, 66, 118, 266
Hydro One, 17, 23, 29-30, 33, 35, 40, 42, 51-53, 58, 60, 62, 66, 71, 75, 77, 81, 85, 95, 99, 105, 107, 109, 117, 125, 133, 197, 211, 225, 235-237, 246-247, 261, 280, 282
Hydro Québec, 172, 189, 250, 252, 254, 283
Hydro tent, 62

Hydro-Electric Extension Fund, 65
Hydro-Electric Inquiry Commission, 79
Hydro-Electric Power Commission, 16, 39, 44, 69, 100
Hydro-Electric Radial Union of Western Ontario, 83
Hydro-Electric Railway Act, 83
Hydro's Financial Restructuring Group, 188
Hydroelectric stations, 47, 179, 192, 216
Hydroelectricity, 98, 110, 120, 210, 254
Hydrogen economy, 256, 284
Hydrogen fuel, 225, 254, 256

IBM, 240
Idaho, 204
IJC, 99-100
Illinois, 89, 170, 173, 184, 249
IMO, 197, 279
Independent Electricity Market Operator, 155, 182, 197
Ingersoll, ON, 44, 46, 53
Insull, Samuel, 89-90, 103, 112, 164, 172, 274
International Joint Commission, 99
Iowa, 184
Israel, 164

James Bay, 224
Japan, 122
Johnson, Tom, 229

Kaministiquia Power Company, 71
Kane, Gregory, 194
Kansas City, 232
Kelsey, Jane, 277
Kent, Edith, 117
Kentucky, 163, 166, 168
Kincardine, ON, 156
King City, ON, 86
King, William Lyon Mackenzie, 50

Kingston, ON, 90, 98, 284
Kingstone, Janet, 117
Kitchener, ON, 30, 32, 44, 50, 266
Klein, Ralph, 202, 210, 234
Kucinich, Dennis, 229-231, 281
Kupsis, Allan, 193

Labour Party, 157
Lake Erie, 24, 72, 216, 267
Lake Huron, 66, 119, 123, 156
Lake Nipigon, 72
Lake Ontario, 24, 27, 72, 98, 119, 123, 125, 127, 133, 267
Lake Simcoe, 56
Langmuir, John Woodburn, 22
Lay, Ken, 86
League for Industrial Democracy, 93
Legislative Assembly of Ontario, 35
Lessard, Wayne, 195
Lethbridge, AB, 65, 234
Lethbridge, John G., 65
Lewis, Stephen, 128, 130
Liberal Party, 17, 19, 29, 31-37, 43, 49, 64, 101-102, 104, 106, 110, 113-115, 125-126, 128-129, 131, 133-137, 139-142, 148, 154, 201, 205, 208-210, 212, 235, 238, 243, 245, 249, 254, 269, 272, 278, 280
Lieutenant Governor-in-Council, 42
Lindgren, April, 280
London, ON, 32-34, 44, 46, 53, 56, 62, 65, 80, 86, 90, 99, 122, 131, 154, 157, 265, 280
London Electricity, 157
London Transport, 154
Long Island, NY, 232
Long Island Lighting Company, 232
Long Island Power Authority, 232
Longstaff, Bev, 235
Lord Dufferin, 22, 263
Los Angeles, CA, 87, 172-173, 232-233, 239

Louisiana, 176, 201
Lovins, Amory, 150
Lyon, T. Stewart, 104

Macdonald Committee, 191-192, 278
Macdonald Report, 192
Macdonald, John Grant, 22
MacDonald, Alex, 129, 150, 255
MacDonald, Donald C., 111-112, 128
Macdonald, Donald S., 190
Mackenzie, William, 26, 237, 263
Mackenzie Syndicate, 26, 28, 33, 56, 70, 73, 82, 84
Magrath, Charles A., 100
Maine, 173, 232
Manitoba, 132, 221, 224, 233, 239, 254
Manitoba Hydro, 172, 189, 239, 250
Manitou Falls, 109
Manufacturers' Life Insurance Company, 33
Market Transition Credits, 203
Markham, ON, 284
Marshall, Stanley, 205
Maryland, 173, 176, 201
Massachusetts, 66, 174-175, 183, 201, 249
Massachusetts Electric, 156
McCarthy, Callum, 158
McGill University, 121
McGill-Queen's University Press, 284
McGuinty, Dalton, 209-210, 254, 269
McKeough, Darcy, 124, 190, 271
McLeod, Lyn, 141, 152
McNaught, W.K., 29, 263
McNeill, Pat, 200, 279
Meaford, ON, 70
Meitner, Lise, 122
Memphis, TN, 232
Michigan, 99, 166, 170, 173, 178, 184, 263

Micro-cogeneration, 255
Microsoft, 240
Middle East, 164-165
Midland, ON, 70
Miller, Frank, 130
Ministry of Energy, 133, 193, 280
   Minister of Energy, 46, 137, 141, 190, 282
Mississagi River, 237, 282
Mitchell, Alanna, 283
Mittelstaedt, Martin, 278
Montana, 16, 163, 201, 204
Montana Power, 177-178
Montreal, QC, 23, 100-101, 121
Moody's Investor Services, 169
Morgan, J.P., 88, 90, 172
The Morgan Group, 89
Moses, Robert, 96
MRI, 257-258
Mulroney, Brian, 115, 140-141
   Mulroney government, 140
Municipal Electric Association, 79, 83
Muny Light, 229-231
Murchison, Clinton Williams, 112
Murphy, John, 193, 197

Nader, Ralph, 156
National Electric Light Association, 86, 88, 104
National Grange, 93
National Grid, 155-156
National Power, 155-156
National Public Ownership Conference, 93
National Research X-perimental, 122
Natural gas, 109-112, 114-116, 118-120, 127-128, 135, 143, 145, 157, 159, 181, 203-204, 211-212, 224-225, 249, 255, 269, 274, 279
   Natural gas-fired cogeneration stations, 137, 221

Natural hydroelectric power resources, 95
Natural Resources Canada, 270
NELA, 86, 88-89, 93
NERC, 175, 200
NETA, 158-160
Netherlands, 250
Network Rail, 161
Nevada, 92, 176-177
Nevius, David, 176
New Deal, 9, 59, 66-67, 87, 169
New Democratic Party (NDP), 19, 47, 62, 97, 112, 119-120, 125-132, 135-137, 139-142, 145-146, 148-149, 151, 157, 182, 188-189, 195, 197, 201, 205-206, 208, 220, 238, 240, 247-252, 270-273, 277-278, 280
   NDP energy policy, 130, 270
   NDP environmental initiatives, 146
   Ontario New Democratic Party (ONDP), 13, 112, 126, 128, 140, 270, 283
New Democrats, 106, 135, 147, 187, 234, 241, 261
New Electricity Trading Arrangements, 158
New England Electrical System, 156
Newfoundland, 31
New Hamburg, ON, 44, 46-47, 53
New Hampshire, 173
New Mexico, 173, 177
New Orleans, 263
New York, 88-89, 92-96, 99, 163, 165, 168, 170, 178, 183, 189, 201, 224, 232, 249, 268, 276, 281, 284
New York City, NY, 22
New York Power Authority (NYPA), 93-94, 189, 268
New York's Consolidated Edison and Public Service of New Jersey, 89

New Zealand, 185, 276-277
Niagara Falls, ON, 21-23, 31, 33, 35-36, 62, 77, 82, 86, 192, 211, 263, 271
Niagara Falls Power Company, 24-25
Niagara generating stations, 192, 278, 280
Niagara Park Commission, 34
Niagara River, 22, 24-25, 31, 73-74, 100, 267
Nipigon River, 71
Nixon, Richard, 165
Norris, George, 90
Norris-Rayburn Act, 66
Nortel, 240, 246
North American Free Trade Agreement, 140
North American grid, 226, 281
North Carolina's Duke Energy, 89
North Dakota, 204
North Sea, 157, 159, 274
Northern Ireland Electricity, 274
Northern Ontario, 67, 71-72, 131, 140
Northern Ontario Pipe Line Corporation, 113
Norway, 122
NRU, 123
NRX, 122-123
NRX accident, 123
NRX reactor, 122
Nuclear Electric Ltd., 155
Nuclear energy, 121-122, 130, 159
Nuclear plant construction, 126
Nuclear reactors, 123, 126, 156, 200
Nuclear station, 123, 127, 129-130, 134, 142, 159, 190, 201, 217, 237, 271
Nuclear Ontario, 10, 121, 252

October 11, 1910 (switching on ceremony), 50-53
Ohio, 170, 173, 184, 230, 249
Ohio Senate, 231
OHIP, 131
OHN, 193-194
Oil, 111-112, 116, 143, 163-169, 211, 254, 274, 279, 283
Oil-burning furnaces, 164
Oil-fired generation, 165, 169
Oil-fired plants, 164
Oklahoma, 173
Omaha, NB, 232
Ontario Conservative Party, 103
Ontario Electricity Coalition (OEC), 283
Ontario Energy Board (OEB), 119, 192, 197, 206, 251
Ontario Grid, 225
Ontario Hydro Affairs, 131
Ontario Hydro Nuclear Affairs, 194
Ontario Hydro Retail, 192
Ontario Hydro's Construction Division, 119, 198
Ontario Hydro's culture, 97
Ontario Hydro's nuclear division, 193
Ontario Hydro's nuclear safety, 134
Ontario Legislature, 22, 25, 30, 83, 200, 251, 263
Ontario Liberals, 209
Ontario Municipal Act, 36
Ontario Power Commission, 36, 264
Ontario Power Company, 26, 69-70, 101
Ontario Power Generation (OPG), 159, 170, 189, 196, 201, 210, 237, 277, 279
Ontario Superior Court, 236
Oregon, 94, 156, 180, 232
Organization of Petroleum Exporting Countries (OPEC), 164
Orillia, ON, 208
Orlando, FL, 232
Oshawa, ON, 70, 85, 269

Ottawa, ON, 22-24, 31, 49-50, 65, 83, 100-101, 109, 112-113, 122, 141, 265, 280
Ottawa Electric Light Company, 23
Ottawa River, 100
Ottaway, Lillian, 51
Owen Sound, ON, 70-71

Pacificorp, 156
Pacific Gas and Electric, 179, 221, 267
Panama Canal, 69, 74
Paris, ON, 44, 46, 91, 99, 122
Paris Group, 122
Peco Energy, 182
Pellatt, Sir Henry Mill, 26, 47, 85, 90, 263
Penetanguishene, ON, 70
Pennsylvania, 24, 31, 123, 156, 163, 166, 168, 170, 173, 176, 182, 201
Pennsylvania Public Utility Commission, 183
Perk, Ralph, 229
Peterborough, ON, 23, 70, 191
Peterson, David, 131, 135, 139
PGE, 179
Phoenix, 87, 232
Pickering, ON, 123-125, 127-129, 133, 136, 193-194, 196, 199-200, 220, 250
Pickering Nuclear Generating Station, 125
  Pickering A, 123-124, 127, 136, 194, 196, 200, 220, 250
  Pickering B, 123, 127
Pickering nuclear rehabilitation, 199
Pickersgill, Jack, 115
Pine Portage, 109
Plebiscite (Ontario), 1923, 84
Popowsky, Irwin, 183
Port Albert, ON, 66
Port Arthur, ON, 55, 71
Port Hope, ON, 122

Porter, Arthur, 128
Porter, Dana, 111
Portland, OR, 94, 268
Portland General Electric, 232
Power Agreement, 1905, 13
Power Circus, 266
Power Commission Act, 82
Power Commission Amendment Act, 49-50
Powergen, 155-156, 158, 160
Power Workers' Union, 14, 149, 172, 191, 193, 197-198, 266
Preston, Gene, 196
Preston, ON, 32, 44, 46, 53, 196
Price caps, 174, 179, 180, 182-183, 203-204, 210-211, 221, 247
Price signals, 175, 180, 214, 218, 220, 275
  Price Signal theory, 218-219, 222
Price spikes, 174-175, 200
Privatization, 15, 17-19, 25-26, 41, 86, 110, 117, 130, 135, 152-161, 163, 170, 178, 185, 188-192, 195, 197-198, 201, 204-205, 207-210, 215, 229-230, 232-237, 240, 243, 248, 254, 267, 269, 274, 276-278, 280-281, 283
  Privatization of public assets, 26
  Privatization of public resources, 232
  Privatization of public utilities, 232
Public Ownership League, 93
Public Power Bus, 206-208
Public Utilities Commission, 180-181, 251-252, 266
Public Utility Holding Company Act (PUHCA) of 1935, 91
Public Utility Regulatory Policies Act (PURPA), 165, 168
PWU, 191-193, 195, 197, 278

Québec, 99-105, 110, 132, 172, 189, 221, 224, 233, 250, 252, 254, 273, 278, 283
Queen Elizabeth II, 95-96, 109
Queenston, ON, 73, 79, 91
Queenston-Chippawa, 69-70, 72, 74-77, 79-81, 87, 106, 212, 278

Rae, Bob, 97, 126, 131, 136, 139-140, 272, 285
Rainy Lake, ON, 13, 225
Rainy River, ON, 135
REA, 66, 197, 267
Reagan, Ronald, 167
REC, 155, 158
Red Deer, AB, 234
Referenda (Ontario)
    1907, 46
    1908, 47
    1913, 62
    1917, 71
Referendum vote, 44, 71, 84, 230
Regent Park, 219
Regional electricity companies, 155
Reiber, Bill, 58
Renewable Portfolio Standards, 250
Renfrew fair (1913), 62
Reno, Bill, 278
Republican Party, 90
Rhode Island, 174, 184, 275
Rihbany, Dave, 208
Rio Algom mines, 124
Rio de Janeiro, Brazil, 149
Robert Moses Generating Station, 96
Robert H. Saunders Generating Station, 95
Roberts, Wallace, 276
Robson, Keith, 284
Rolls Royce, 154
Rolphton, 123, 127
Roman, Stephen, 125-126
Roosevelt Administration, 91
Roosevelt, Franklin D., 9, 59, 87, 91, 93, 95, 100, 275
Roosevelt, Theodore, 94
Rosenberg, Bill, 277
Ross, George, 29
Ross Power Act, 36, 39
Rothman, Mitch, 274
Royal Commission, 79, 84, 102-104, 115, 128, 285
Rumpel, Hilda, 52, 266
Rural Electrification Administration, 66, 91, 93, 267
    Rural electrification campaign, 60-61
Rush, Eby, 58

Sacramento, CA, 232, 276
Sagall, Sabby, 274
Saginaw River, 263
St. Jacobs, ON, 30, 32
St. Lawrence system, 55, 82, 98-101, 107
St. Lawrence River, 94-96, 98-101
St. Lawrence Seaway, 99-100
St. Mary's, ON, 44, 46, 53
St. Thomas, ON, 44, 46, 53, 55, 62, 102
San Antonio, TX, 232
San Diego, CA, (Gas & Electric), 179
San Francisco, CA, 286
Saskatchewan, 114, 204, 224, 233
Saudi Arabia, 169
Sault Ste. Marie, ON, 71, 206, 237
Saunders, Robert, 96
Savings & Loan, 167
Schaefer, Dan, 172
Scott, Helene, 206
Scottish Nuclear, 155
Scottish Power, 156, 274
Seattle, Wash., 87, 180, 232
Securities and Exchange Commission (SEC), 91, 184, 268
Select Committee, 125, 128-129, 131-132, 194, 271

Severn River, 70, 73
Sharp, Mitchell, 113
Shell, 254, 279
Shortage of generation, 182
Shortage of power, 178
Siemens Electric, 190
Silver Falls, 109
Silverado Banking, 275
Simcoe Light and Power Company, 70
Sir Adam Beck Generating Station, 76, 261
Small-scale generation, 226
   Small-scale green power developments, 146
Smith, Alfred E., 94
Smith, Cecil B., 43, 46
Smithsonian Institute, 86, 274, 285
Snider Commission, 39, 41, 43
Snider, E.W.B., 30
Social Contract, 151
Solar cells, 18, 137
   Solar electric generators, 255
   Solar electricity panels, 67
South America, 153
Southern Company, 89, 157
Southern, Nancy, 204
Special Committee of Inquiry, 102
Spence, F.S., 28-29
Stanford University, 110
State of Victoria, 277
Steel Company of Canada, 55
Sterling, Norm, 192
Stevenson, James, 279
Strategic and Investment Planning, 200
Stratford, ON, 44, 46, 53
Stockwell, Chris, 205
Strong, Maurice, 149-151, 187, 273
Stouffville, ON, 85
Sudbury, ON, 237
Sutherland Report, 84
Sutherland, Sylvia, 191-192
Sweezey, R.O., 103

Switzerland, 29
Syracuse University, 92, 94, 101

Tennessee Valley Authority, 90-91, 93, 189, 196, 232-233, 268
Tesla, Nikola, 22, 263
Thatcher, Margaret, 154
Thompson, Carl D., 93
Three Mile Island, 156, 166, 194, 230
Thunder Bay, ON, 55, 71-72, 115
Tim Hortons, 208
Toronto, ON, 18, 23-24, 26, 28-29, 31-32, 34-35, 41-48, 50, 53, 55-56, 70, 75, 79, 83-85, 99, 101, 123, 125, 135, 148, 178, 191, 193, 200, 219, 225, 246, 253-255, 263, 268-270, 272, 277-279, 281-282
   Toronto City Council, 29, 31, 84, 268
   Toronto Junction, 44
   Old City Hall, 41-42
Toronto Blue Jays, 272
Toronto Electric Club, 200, 279
Toronto Electric Light Company, 23, 26, 28-29, 41, 85
Toronto Hydro, 75, 253
Toronto Renewable Energy Cooperative, 225, 253
Toronto Street Railway, 26, 41, 85
Toronto Transit Commission (TTC), 84
TransCanada PipeLines (TCPL), 112-114, 190
Transmission grid, 82, 101, 225-226, 246
TXU Energy, 157

U.S. Congress, 95, 201, 268
U.S. Department of Energy, 173, 276
U.S. deregulated jurisdictions, 220
U.S. deregulation, 163, 165, 171, 200
U.S. Federal Trade Commission, 89

U.S. FERC regulations, 272
U.S. Northeast blackout, 1965, 175, 276
Ukraine, 134
Union of Canadian Manufacturers, 31
Union Gas Ltd., 191
United Auto Workers, 269
United Farmers of Ontario, 62, 64, 81
United Kingdom, 10, 31, 49, 86, 90, 121-122, 153, 156-157, 159, 161, 163, 176, 224, 250, 265, 274
United Nations Framework Convention on Climate Change, 149
University of Toronto, 75, 191, 284
Urquhart (Toronto mayor), 35
Utilicorp, 235

Vancouver, BC, 284
Victoria Hospital, 122
Virtual Power Plant, 255

Wales, 32, 157, 274
Walker, Jim, 58
Walker, Sir Edmund, 48
Walkerton, ON, 146
Walper House, 32
Warner, Burt, 58
Wasdell Falls, 70
Washington D.C., 86, 99, 173, 175, 180, 231, 268, 276, 285
Washington State, 92, 234
Water Boundary Treaty, 99
Waterloo, ON, 32, 44, 46, 53, 58, 62
Waverman, Leonard, 191
Wawa, ON, 206, 208
Welland Canal, 46, 99
Welland River, 74
Wentworth County, ON, 65
Western Electric, 240
Weston, ON, 44
Whent, Howard, 208

Whitedog Falls, 109
Whitney, James Pliny, 35, 37
Wildman, Bud, 147
Williams, G. Mennen, 99
Wilson, Edward O., 147
Wilson, Jim, 195-197, 201, 205, 235, 270
Wilson, Pete, 172, 205
Windshare, 253-254
Wind turbines, 67, 137, 250
Windsor, ON, 55, 62-63, 279
    Windsor City Council, 62
Winnipeg General Strike, 64
Wisconsin, 93
Wood Gundy, 191
Woodbridge, ON, 208
Woodstock, ON, 44, 46, 53, 62
World War I, 55, 63, 70, 82
World War II, 100, 105, 121, 154
WorldCom, 41, 246

AGMV Marquis
MEMBRE DE SCABRINI MEDIA
Québec, Canada
2003